WHEN FOOD KILLS

WHEN FOOD KILLS

bse, *e. coli*, and disaster science

T. HUGH PENNINGTON

OXFORD

UNIVERSITY PRESS

OXFORD
UNIVERSITY PRESS

Great Clarendon Street, Oxford OX2 6DP

Oxford University Press is a department of the University of Oxford.
It furthers the University's objective of excellence in research, scholarship,
and education by publishing worldwide in

Oxford New York

Auckland Bangkok Buenos Aires Cape Town Chennai
Dar es Salaam Delhi Hong Kong Istanbul Karachi Kolkata
Kuala Lumpur Madrid Melbourne Mexico City Mumbai Nairobi
São Paulo Shanghai Taipei Tokyo Toronto

Oxford is a registered trade mark of Oxford University Press
in the UK and in certain other countries

Published in the United States
by Oxford University Press Inc., New York

A catalogue record for this title is available from the British Library

Library of Congress Cataloging in Publication Data
(Data available)
ISBN 0-19-8525176

10 9 8 7 6 5 4 3 2 1

Typeset by Newgen Imaging Systems (P) Ltd., Chennai, India
Printed in Great Britain
on acid-free paper by
Biddles Ltd., Guildford & King's Lynn

Contents

Preface *vi*
Acknowledgements *viii*

1   *E. Coli* O157, Central Scotland 1996 *1*

2   Why Disasters Happen *25*

3   Unlearned Lessons *41*

4   The Inspectors Fail *58*

5   Inspectorates have Limits *65*

6   *E. Coli* O157 *96*

7   Other *E. Coli* *103*

8   CJD *115*

9   The Science of TSEs *124*

10   BSE *143*

11   BSE—Why Things went Wrong *172*

12   vCJD—The Future *185*

13   The Precautionary Principle *192*

14   BSE, vCJD, and *E. Coli*: The Aftermath *198*

Further Reading *214*

Index *221*

Preface

William Shakespeare has a lot to answer for. The
'Macbeth' image of Scotland is still powerfully influential.
It is a story of treachery, murder, guilt and revenge, set in
the olden days of castles, forests, and private armies. But
while it may bring tourists and inspire artists and musi-
cians its relevance to present-day life must, surely, be
utterly remote. After all Scotland is a modern industrial-
ized country where monarchs no longer kill their way to
the throne. Nevertheless assassins are still at work. They
are agents like *E. coli* O157 and BSE. And the circumstances that aid them are as dramatic and relevant to the
human condition as the story told by the Scottish play.
Like many of Shakespeare's other works they have strong
international backgrounds. There is still a particularly
Caledonian component to their doings, however. *E. coli*
O157 and vCJD, the human form of BSE, are commoner
in Scotland than in any other country, and many of the
most important scientific facts about the disease-causing
agents in the BSE family were discovered there.

Lady Macbeth herself also points to another particularly paradoxical Scottish connexion. A continual theme that runs through any account of food poisoning is handwashing. Its underperformance helps food poisoning organisms to travel through human communities better than anything else. After declaiming 'out damned spot! out, I say!' and 'all the perfumes of Arabia will not sweeten this little hand' she exclaimed 'wash your hands'. If only food handlers followed her example! The paradox is that handwashing as an anti-infective measure was invented in Scotland by Joseph Lister. His development of antiseptic surgery in Glasgow in the 1860s was one of the most important and revolutionary developments in medicine. Getting rid of bacteria from the surgeons hands was one of its central elements. This book explains why the Scots have been so bad at following Lister's—and Lady Macbeth's—example and describes the disasters caused by its neglect.

But it is not just about Scotland. BSE is above all an English disease. It started there and was one of its more successful late twentieth-century exports. Sir Andrew Aguecheek and Sir Toby Belch, although nominally 'Illyrians', are as English as any of Shakespeare's characters. The Playwrights' prescience in 'Twelfth Night' was remarkable—again—as he has Sir Andrew saying 'but I am a great eater of beef and I believe that does harm to my wits', and Sir Toby replying, 'No question'.

Acknowledgements

Whatever Britain's failings in food safety, it cannot be said that those working to make things better are unhelpful or unwilling to discuss problems and solutions. Without their help and that of the many institutions to which they belong this book would not have been possible. In particular, I thank the members of the Central Scotland *E.coli* O157 Outbreak Report Group and its Scottish Office officials. I have received the strongest support over the years from my departmental colleagues at Aberdeen University and from its Principal, C. Duncan Rice, for which I am enormously grateful. I also thank Ian Sherman, and Catherine Pennington, Wilma Odell, Eve Seguin and Ian McCann for reading the manuscript—any errors are mine—and Alan Dickinson and Nicola Hepburn for providing photographs. For permission to publish illustrations, I acknowledge the Press Association (Figures 1.1 and 10.2), Hulton Archive (Figures 2.2, 2.6, 5.2 and 11.1), Her Majesty's Stationery Office (Figures 2.4 and 2.5), Aberdeen Journals (Figure 3.1), the South West

Yorkshire Mental Health NHS Trust (Figure 3.2), the Western Mail (Figure 5.1), Bildagentur Huber (Figure 5.5), the National Portrait Gallery (Figure 5.8), Dr A. Holly Dolin (Figure 5.9) and the International Solvay Institutes for Physics and Chemistry (Figure 9.1).

Without Carol this book would not have been written. The contribution of my secretary, Jacqueline Morrison, has been outstanding throughout. My gratitude to both is immense.

Chapter 1

E. coli O157, Central Scotland 1996

Wishaw is a medium-sized town near Glasgow. It owes its existence to coal and iron. In the middle of the nineteenth century it was one of the most prosperous places in the world. Economic migrants came from the Highlands, England, Ireland, Poland, and the Baltic states. The failure of Scottish ironmasters to keep up with technology and the exhaustion of coal seams meant that the glory days did not last, and for the last century it has been a much more inward-looking place. The main legacies of immigration are strange surnames and Orange Lodges. In 1996 its citizens were served by two hospitals equidistant from the town. Monklands General, opened in 1976, had an infectious diseases unit. However, most patients went to Law Hospital. This was built at the beginning of the Second World War with 1000 beds to cope with air-raid casualties. At the end of the war it transferred to civilian use. Events in its laboratory in November 1996 were to put Wishaw on the map. Just as Glencoe is remembered for its massacre and Culloden for the defeat of bonny Prince Charlie, Wishaw will forever be associated with its lethal outbreak of *E. coli* O157.

The sample that signalled its beginning had been passed by a little girl—a 5-year-old—on Wednesday, 20 November 1996. It was a stool. This word carries the impression of firmness, even of a deliberate effort in its production. Hers was not. It had been sent to the laboratory by her general practitioner because it had no form at all. She had diarrhoea. It was dealt with by the laboratory staff in the normal routine way. Along with urine, spit, pus, and blood it was part of their daily bread and butter. But the result that came through on the Thursday was not routine. It looked as though their diagnosis was going to be an *E. coli* O157 infection. If confirmed, it would be only the second case of the year at Law. Kenneth Liddell ran the bacteriology diagnostic laboratory. He had met this organism before, so he needed no reminding that it caused extremely nasty infections. On the basis of a still provisional result he took action. He telephoned to warn the girl's general practitioner that her patient might be infected with a particularly dangerous bacterium. He started detective work to find its source by asking whether certain foods had been eaten by the girl—in particular the kinds known to have transmitted infection in other outbreaks. Nothing came up that caused any concern. So it was reasonable to assume that the case was a single, sporadic event—as most *E. coli* O157 infections are. But everything changed the next day, Friday, 22 November. The girl's *E. coli* O157 provisional diagnosis was confirmed—and there were two more provisional positives, one from a 14-year-old boy and another from a 60-year-old patient in one of the women's wards in the hospital. All three patients had addresses in Wishaw. It was still possible that all this was coincidental. Liddell tested this idea. He trawled through the data that had come with all the faecal samples that had been sent to his laboratory the previous day. None had grown *E. coli* O157. But two were from patients that had had bloody diarrhoea. This substantially raised the odds that they also had the infection. Both had addresses in Wishaw. More detective work followed. But it was not enough just to try and track the source of infection. Direct action had to be taken to stop it spreading in the hospital. Liddell's own experience with O157 had shown him how easily this could happen. In 1990 he had investigated an outbreak caused by it at Hartwoodhill Hospital. This was a small psychogeriatric institution a few miles from Wishaw where a patient with diarrhoea infected seven other residents and three staff. Four of the patients died. Hygiene standards at the hospital were not high. The sheriff who conducted the fatal accident inquiry into the deaths (the Scottish equivalent of the English coroners' inquest) concluded that a control of infection nurse 'could reasonably have been expected to ensure that the justifiable criticism of lax hygiene in the uplifting of soiled laundry, of unsafe standards of cleanliness in the wards and their furnishings, and of poor standards of housekeeping in the washing up of cleaning bowls and utensils would not have been levied'. All that Liddell had found at Hartwoodhill in the hospital's favour was that two cockroaches caught in its kitchen were negative for *E. coli* O157.

Faced on that Friday morning with one definite, two likely, and two possible cases it was time to leave the laboratory and see what was going on in the hospital. In any event other important things were happening out there as well. There was a meningitis case in the paediatric ward. Liddell went there, and dealt with it. He was told that the 14-year-old had been transferred to the male surgical ward so he went there to interview him and his mother. He was already in isolation because of his bloody diarrhoea. He asked about the exposure he had had, what he had eaten over the past few days, and whether he had had anything different from the rest of the family, who had been unaffected. He did the same with the other two patients. One was in considerable pain. Two potential sources emerged from these interviews. First of all there were reports from three people of meat or sandwiches containing meat having been bought from a butcher's called 'John Barr'. The second was that two, or possibly three, had purchased milk from a local supermarket. Liddell widened the net of his inquiries. He phoned Andrew Todd, the Infectious Diseases Consultant at Monklands Hospital, and asked him whether he was aware of any *E. coli* O157 cases he had been sent or had been informed of in the past few days. He said he had two cases of bloody diarrhoea in his Unit, both from the Wishaw area. He contacted all the Wishaw general practitioners. One had some very interesting information. He had seen five patients with bloody diarrhoea, three of whom had been at a lunch in Wishaw Parish Church the previous Sunday. The other two patients he mentioned were from the community. One of the patients was a diarrhoea case that so far had negative bacteriology results. He went back to the laboratory and looked again at the plate on which her culture had been grown. After an intensive search he found one suspicious colony. He rang the senior local public health specialist, Syed Ahmed, to tell him what was going on, and met him at 4 PM.

It was clearly necessary to set up an Outbreak Control Team, and it was agreed that this should be done and that its first formal meeting would be arranged for 11 AM the following day. But the rapidly evolving situation, and the loose ends that needed rapid resolution—particularly the relative importance of the butchers shop versus the milk supply as the sources of infection—meant that it was obvious that doing nothing until then would be wrong. A meeting between Ahmed and his public health colleagues Martin Donaghy and Helen Irvine, with Graham Bryceland and Jeffrey Tonner from the North Lanarkshire

Environmental Health Department was called for 6 PM. Although the evidence that milk was the culprit was not particularly strong, no-one could forget the largest *E. coli* O157 outbreak that had previously happened in Scotland. It centred on a rural community in West Lothian near Bathgate, a town less than 20 miles from Wishaw and had occurred only $2\frac{1}{2}$ years before. One very young child died early in the outbreak. Of the 71 other cases nearly half were aged 5 or under, and nearly a quarter of these young children had kidney complications. Altogether eight children and adolescents needed artificial kidney support. Three of them lost their kidneys and needed kidney transplants and the kidneys of another two were permanently damaged. The outbreak was caused by contaminated milk sold by a small dairy which got its supplies from feeder farms and then pasteurized, bottled, and delivered it. Although it was not possible to establish whether the fault lay in defective pasteurization or later contamination at bottling, or both, the dairy proprietors, R. and M. Haston, were found guilty at a criminal trial of selling contaminated milk. So at the six o'clock meeting in 1996 it is not surprising that milk remained on the agenda. Nevertheless it was clear that a strong case was building that pointed to John Barr's butchers shop as a likely source. But nobody at the meeting had heard of him, far less had any detailed knowledge about his business. A point in his favour was that he was not known to the senior Environmental Health officials as a problem. Also in his favour was that butchers had not previously had any noteworthy association with *E. coli* O157 infections.

Nevertheless, it was decided to visit him that evening. By this time 15 likely *E. coli* O157 victims had been identified, eight with a link to Barr's. Graham Bryceland had to leave the meeting early to go back to his head office but on his way he went via Wishaw and stopped to look at Barr's through the window. It had already closed. He was impressed. The shop looked neat and tidy. He also saw that there was a Barr bakers shop next to the butchers. He was not alone in liking the look of Barr's. John Barr was a member of his trade association, the Scottish Federation of Meat Traders Association. In a competition open to its members, customers had voted him 1996 'Scottish Butcher of the Year' on the basis of front shop appearance, friendliness of service, and range of products. Bryceland went to the police to find out where John Barr lived, and met up with Jeffrey Tonner and Syed Ahmed at Barr's house at 9 PM. They were invited in. He was told of the suspicion that cooked meats from his premises had a problem, and that because this was *E. coli* O157 it was a serious one.

John Barr was told about the West Lothian outbreak to impress on him its bad effects. Graham Bryceland asked him to stop selling cooked meats voluntarily. He agreed at once, but asked whether the bakery shop run by his son next door could go on selling sausage rolls and bridies. Christmas was coming and these were particularly important for the bakery business. He assured the group that it operated quite separately from the butchery side, and because of this it was agreed that these things could go on being sold. It was arranged that Environmental Health Officers would arrive at the business at 7 AM the next morning to remove all suspect food and start a full investigation. The last meeting that night was between John Barr, his two senior managers, and his son. He told them that there had been a food scare, about his voluntary agreement to stop selling cooked meats, and that the Environmental Health Officers were coming the next day. They planned their reaction.

Any complacency that all that had been done was all that was necessary, or that things were under control, was comprehensively dashed first thing on Saturday morning. Overnight the number of patients had doubled. They were queueing up to be admitted to hospital. The hospitals themselves were beginning to be seriously stressed. To make things worse the environmental health officers came back from John Barr's to the 11 AM meeting of the Outbreak Control Team with bad news. Their interviews with staff had indicated that there could be cross-contamination between the butchers and the bakers. There were concerns about the hygiene practices of some of the staff. But to further muddy the waters it appeared that some of the new cases did not seem to have any links to Barr's. They did not live in Wishaw. However, there was some information suggesting that Barr's supplied other shops. So was there one outbreak—or two? The 11 AM meeting finished about 2 o'clock. It was decided to ask John Barr to stop selling all products not requiring further cooking, and it was agreed that a public statement would be made about what was going on, including the identification of his premises. Barr agreed to these actions. The Press Release finally went out at 5.30 PM. It was delayed because British Telecom had lost the Help Line telephone number for Monklands Hospital and needed time to find it again. The outbreak was top of the evening TV news.

Barr's had been busy that day. Martin Barr and the other bakers started work between 3 and 4 AM. John Barr and staff on the butchery side were in by 6.30 AM. Some cleaning was done and cooked meats were taken out the cold store and vacuum-packed ready to be taken away by the Environmental Health Officers. Just before 7 AM they arrived at the shop and were let in. Jeffrey Tonner and Drew McLean soon realized that they had a big job to do. The premises were much larger than they expected. They had planned to interview half the staff before the 11 AM meeting—but did not expect to find that 40 people worked there. They also had to take swabs and samples of meat for bacteriology tests.

Jeffrey Tonner graphically described what he found when he was questioned 18 months later by the Procurator Fiscal Depute at the Fatal Accident Inquiry.

> I formed the view from interviewing people there that the food hygiene practices within the premises were pretty poor.
>
> The food hygiene practices were poor?—Yes
>
> In what way were the food hygiene practices poor?—The young lad was going from cutting and preparing raw meat to cutting and preparing cooked meat without washing his hands, and he was using the same knife, so if there was anything on the raw meat it was going to cross-contaminate the cooked meat. I didn't like that much.
>
> Was he quite open about this?—Oh, yes. All the staff in my judgement whom I interviewed, or the vast bulk of them, were completely open.
>
> When you talked about the young man going from raw to cooked meat and using the same knife with unwashed hands, did you see him doing that?—No, he told me he did it.
>
> You asked him?—Yes.
>
> And he told you that?—Yes.
>
> Can I explore this a bit further? What type of questions did you ask that elicited that answer from him?—I just asked 'What is your job? What do you do?', and he gave me an answer that he was in the butchery, and he was cutting raw beef etc. I said 'Do you ever cut cooked meat? Do you deal with cooked meat?' and he said 'Yes'. I said 'Do you wash your hands between dealing with raw meat and cooked meat?' and he said 'No'. I said 'Do you clean or wash the knife used for cutting the raw meat between it cutting the raw meat and the cooked meat?' and he said 'No'.
>
> Did he seem to have any idea of the risks which might be involved when you were speaking to him?—No.
>
> I think you were going to mention other poor practices which became apparent to you that morning?—The lady handled raw meat and then handled cooked meat, and she didn't wash her hands. I asked her about that, and she said she didn't.
>
> Was the lady a front shop employee of the butcher's—Yes.
>
> I spoke to Mr Hepburn. He is the factory manager. Do you recall when it was you spoke to him? Was it on the morning when you were in the premises?—Yes, that was all in the morning. I interviewed him in the morning.
>
> So did something cause you some concern when you spoke to Mr Hepburn?—Yes, Mr Hepburn was...he seemed to be aware...he didn't give me a great deal of cause for concern regarding the food, his food practices, that could cause cross contamination. I asked him, I said 'What do you clean the premises with?', to clean

the utensils and machinery and he said he would go downstairs and bring it up to show me which he proceeded to do and he showed me it and it was just a liquid detergent. It wasn't a bactericidal detergent which I would have anticipated they would have used. So I said to Mr Hepburn 'That is not a bactericidal cleaner you are using there' and he just looked at me. He didn't know what I was talking about.

How was it you were able to determine that this was a non-bactericidal cleaner?—It is a liquid detergent, fully bio-degradable and if it is a bactericidal cleaner they are more expensive and they always indicate one way or another it is bactericidal. They indicate that is what they do. Because they say it is more expensive, well it is a much more expensive product so they always indicate on it one way or the other. This was just a 5 litre plastic jar with a label on it and that is all the label said so I was a bit concerned there.

Did you take any further action in relation to that matter, the detergent?—Well, time had really pressed on now and Mr McLean and myself really had to be at this meeting at 11 o'clock so we went downstairs and frankly I thought Mr Hepburn had made a mistake and I really wanted to check this with Mr Barr so I just took the time and I approached Mr Barr and said 'You're not using bactericidal cleaner in your factory to clean your factory and the machines' and he said 'Oh, no, we are' and I said 'You're not' and he sent Mr Hepburn to fetch it and he brought it up. He brought a full . . . the other one was just about a quarter full but this time he brought me out a full 5 litre container, a plastic container, of this cleaner, the liquid detergent fully bio-degradable and Mr Barr looked at it and I said 'That is what you are using'.

Mr Barr looked at it and said what?—No, Mr Barr looked at it and I said 'That's what you are using' and he didn't say anything. He didn't say anything at all but he didn't look very happy.

So Mr Barr didn't say anything?—No, he didn't say anything. He just sort of blanched a bit.

The Environmental Health Officers left and Tonner described his findings to the Outbreak Control Team meeting.

At Barr's the voluntary agreement was proving leaky. Cooked meat orders prepared on Friday were still going out to customers. One of them, David Moon, came by in the morning to pick up a birthday cake and a hundred slices each of roast beef and boiled ham and roast turkey for a birthday party he had arranged for his 18-year-old niece at the Cascade Public House at Wishaw that evening. John Barr gave him the order and the girl's mother took the cooked meats home and prepared a buffet of sandwiches, rolls, cooked meat slices with salad, cakes, sausage rolls, and quiche. Moon was told by his niece that there were rumours that John Barr's premises were implicated in an outbreak of food poisoning. He was unaware of these but telephoned the premises and was informed that the cold meats he had just purchased were 'fine'. Content with this information, he reassured her mother and left to go away for the weekend. The rumours linking food poisoning with John Barr's premises got stronger so later in the afternoon she telephoned the Wishaw Police for advice. They referred her to Law Hospital. She telephoned and was referred to a doctor to whom she explained the situation. The doctor did not know anything about any food poisoning outbreak and gave reassurance that everything would be 'OK' and that her party could enjoy their food. So she finished preparing the buffet. Her husband and son delivered the food to the Cascade Public House and the party started later that evening. More than 100 people were there either as family members, guests, public house staff, or gatecrashers. At the end of the party drinkers at the main bar were invited to finish the food.

Later on Saturday John Barr had second thoughts about the Moon order. He tried to get hold of him by telephone, but failed—and left it at that, after leaving a message. Environmental Health Officers came back at about 2.30 PM to take more swabs and to get the addresses of Barr's customers—not over the counter ones but the other business and institutions supplied since the beginning of November. John Barr's wife Elaine came in at 3.30 PM to do this. After going through order books for about an hour she prepared a handwritten list. Jeffrey Tonner was relieved to see that the outlets listed were all local. Even better, all seemed to be small such as the Pather Post Office at Holytown and the Wishaw Golf Club.

Sunday should have been a routine day spent following up actions left incomplete from Saturday. Barr's was closed but many of his staff came in to carry out a 'deep' clean of the premises. Jeffrey Tonner and other Environmental Health colleagues continued their interview of Barr's staff. But just after 1 PM he was given information that shook him and shocked him. With a colleague, Drew McLean, he was interviewing John Holloway. He said that he was a delivery driver and that he delivered raw sausages, about $3\frac{1}{2}$ tonnes a week, as well as cooked meat and pies, to three Scotmid stores in Bonnybridge. This took the area potentially affected north to the Forth Valley—well beyond the boundary of Tonners' local authority, and involved a firm with a large number of branches. It also indicated that the list of customers made by Mrs Barr was seriously defective. The head of the Falkirk Environmental and Consumer Protection Department in Forth Valley was told. Confirmation of this extension of the affected area came quickly the next morning when two cases with clear links to Barr's were identified. Both had eaten cooked meats bought at the

Scotmid Store in Bonnybridge High Street. So Jeffrey Tonner's shock was justified. Information about contaminated food was coming in far too slowly for comfort. But if he had known what was happening in Forth Valley on the Saturday afternoon—at almost exactly the moment when he was reading the list of Barr's customers prepared by Mrs Barr and breathing a sigh of relief because they were all local—his shock would have turned to horror.

At about II AM on the Saturday morning Norma Adam, the Catering Manager of the Bankview Nursing Home—a private high-quality institution specializing in the care of those with the dementia of old age—made a casual purchase of £1.69 worth of cold meats and smoked ham at the Scotmid shop at Kilsyth Road, Haggs, Bonnybridge. She did this because of a breakdown in her normal weekend ordering arrangements. During the previous week the shop had been supplied with John Barr roast beef, roast pork, and gammon. These had been sliced on the single slicing machine in the shop, which was also used to prepare Norma Adams's purchases—which were not John Barr products. When she got back to the nursing home she repacked the meats in cling film and put them into the refrigerator. At lunchtime they were used to make sandwiches, which were given to the Bankview residents in the later afternoon. More sandwiches were prepared from the same meats and eaten by the residents on the next day, Sunday. So at this point it was possible that all the residents—32 permanent and 10 respite care—had been exposed to infection—if the John Barr meats supplied to Scotmid had been contaminated with E. coli O157, and if bacteria had been transferred by the slicing process to the Scotmid products bought by Nora Adams. Confirmation that this had indeed happened began to come through on Wednesday, 27 November when three residents developed diarrhoea.

The Hartwoodhill outbreak had shown how vulnerable the elderly were to an E. coli O157 infection. But the Bankview residents were not the only vulnerable group that had been put at risk by eating the bacterium from Barr's. On Friday afternoon Kenneth Liddell's detective work on bloody diarrhoea cases had already identified three elderly patients who had all attended a lunch at Wishaw Parish Church on the previous Sunday, 17 November. For several years every six months the church had provided a meal after a service for its over 70-year-old members. On this occasion the food had come from Barr's, who were regular suppliers to the church for a weekday café that it ran. During the week before the lunch Mrs Davidson, the minister's wife, and her helpers met to pick the menu. It was decided to serve beef broth, steak pie, and a fruit cocktail with ice cream. Boiling beef for the broth and 30 lb of cooked beef stew and 100 pieces of cooked pastry were ordered. On Friday, 15 November Barr staff loaded two plastic cooking bags each with 15 lb of beef trim. Between noon and 1 PM they were put in a large boiler with another 20 lb of beef. It was switched on and left until 4 PM when it was switched off and left overnight. Early next morning—before 7 AM—the bakers came in and took out the bags. What happened next is not entirely clear but it is likely that they were put into white baskets at the side of the boiler—possibly ones that had been used for the storage of sausage and other raw meats. Along with the pastry, which had been cooked at Barr's to about 200 °C for about 15-20 min and put in a bread basket covered in greaseproof paper, the bags were collected in the late morning by Ronnie Holloway, Barr's other van driver, and delivered to the church hall. When they arrived the minister's wife Mrs Davidson was preparing vegetables for soup. At her request Holloway put the stew bags on chairs in the conference room across the hallway from the kitchen. They were noticeably warm. The pastries arrived on a white plastic board in two layers, the upper separated from the lower by greaseproof paper. The top layer was not covered. The board was also placed on chairs in the conference room separate from the cooker stew. Mrs Davidson put the vegetables into the catering pot containing the broth mix and beef cube stock. She took the boiling beef, which had been delivered last, through to the kitchen, removed it from its wrapping and rinsed it under the cold water tap. She rinsed a pot in boiling water from a boiler, put the beef in the pot together with cold water, put the pot on the cooker, and washed her hands. The boiling beef was left to boil for about an hour, when Mrs Davidson put the stock from the beef into the catering pot with the other ingredients. The contents of that pot were in turn boiled for about an hour.

After washing her hands Mrs Davidson dealt with the stew. She cut the bags open at the top with a knife and poured their contents into two grundy tins—aluminium cooking tins 10 × 15 in. and 3 in. deep. There was an excess of gravy which she decanted into a 4-l ice cream tub, which had been washed out before she used it. She left the stew, and excess gravy, in the conference room. The lids of the grundy tins were left off so that the stew would cool. Before leaving, just after one PM she put the lids on the grundy tins and clean tea towels over the pastry. She took the boiling beef away for the fox that visited the Manse garden.

The stew, pastry, and excess gravy were left overnight in the conference room. She did not refrigerate the stew because she thought that it was wrong to put aluminium containers into a refrigerator. The room was not heated overnight.

At 9 AM on Sunday morning Mr and Mrs Davidson arrived at the church hall. One of them lit the oven. Mrs Davidson went to Safeway to collect bread and any other necessities for the lunch. Mr Davidson went to the church to prepare for the service. Mrs Davidson returned to the church hall at about 9.30 AM. Helpers arrived. They set out the tables and took dishes from a special cupboard in the hallway. Mrs Davidson helped to lay the tables. At about 10 AM Mr Ferrier arrived. The oven was on. The door of the oven was open. Mr Ferrier took the major role in the preparation of the stew and of the meal generally. He had prepared meals on this scale before. He reset the temperature of the oven to 500 °C, its maximum temperature setting. He put the tins of stew in the oven. The potatoes and other vegetables were put on. At about 12.10 PM Mr Ferrier opened the oven door, reset the temperature to 300 °C and loaded the pastry. The church service ended about 12.30 PM. Mr Davidson changed and went into the church hall arriving at about 12.45 PM. He said grace and service of the meal began. The main course was dished out using a production line method. Pastries were taken out 10 or a dozen at a time. Stew was added. Two plates at a time were taken by each helper to the diners. The tins of stew were used one at a time, the unused tin remaining in the oven which remained on at 300 °C. None of the helpers recollected adding gravy from the ice-cream tub to the stew. The soup was piping hot and the rest of the meal was also hot. Before the stew was served it is said that it was beginning to bubble, to be very hot. Steam arose from the potatoes in the pot and those mashing the potatoes took turns to do so because of the heat. The pastries crackled as they came out of the oven. They were really hot.

Even if the staff at Bankview and Mrs Davidson and her team had known that products from John Barr could contain live *E. coli* O157 they should have had little cause for concern. After all the Bankview meats came from another clean, non-Barr source and the only connection with Barr's was that they had been sold by a shop handling Barr products. By all accounts the Wishaw Old Parish Church steak pie had been cooked not once but twice at temperature/time combinations that should have killed the bacterium many times over. But events were to show that if these views had been held their optimistic conclusions would have been cruelly dashed.

On Monday more and more cases came to light and more people continued to fall ill. Clearly the outbreak was going to be big. News had got out that Barr's might be the source of the *E. coli* O157. Despite this he was still trading briskly (Figure 1.1). Photographs of regular customers in his shop and quotes from his supporters that 'this scare is certainly not going to put me off' and 'Poor John is being murdered for it. Things are pointing towards him but we have not got one iota of proof about it' were appearing in the papers. But events

Figure 1.1   Barr's still trading, Saturday, 23 November 1996.

took a step-change for the worse the following day. On Tuesday, 26 November, Henry Shaw died. Mr Shaw was 80 and had been in good health for his age. He was an elder at Wishaw Old Parish Church and had attended the church lunch. He enjoyed the soup but was a poor eater and only had a little of the steak pie. He became ill the following Wednesday. The general practitioner visited and prescribed antibiotics and anti-diarrhoeal medication. Later that day he began to pass blood. On Thursday he was admitted to hospital. A firm diagnosis of *E. coli* O157 infection was made late on Friday. He condition declined and he became very distressed. By Monday he was unconscious and he died at a 6.45 AM the following day.

Mr Shaw was not the only person to die that day. Mrs Annie Criggie had developed symptoms on Friday, 22 November. She lived in a flat at Bonnybridge, and liked cold meat sandwiches. She ate one every day. She bought cold meats from the Scotmid store in the High Street, a shop that sold John Barr products. By Tuesday she was very unwell. Her speech was affected. The general practitioner was called and visited her that morning. She was not dehydrated. Her daughter visited her several times during the day. At about 7 PM she discovered her lying on the floor. She was dead. In her fridge was an opened packet of meat labelled 'John Barr's Roast Pork'.

Even before these two deaths became publicly known, events were moving fast on the outbreak control front. At the Tuesday Outbreak Control Team meeting Graham Bryceland said that the list of Barr outlets—shops supplied by him and so on—was getting much more complicated than he thought on Monday. The numbers were growing. But not only were the investigators discovering new shops supplied by Mr Barr, it was coming to light—in words used at the meeting— 'that they then supplied somebody else and then somebody else supplied somebody else'. The whole situation was becoming very complex, not only locally, but throughout the central belt of Scotland. Immediately after the meeting Syed Ahmed telephoned the Scottish Office and told them 'it may be not only Lanarkshire'. He wanted reassurance that despite Bryceland as the food expert telling him that he had done everything possible that needed to be done, there was nothing else that he should do. He requested a meeting with Stephen Rooke, the Chief Food and Dairy Officer of the Scottish Office. But before he rang the Scottish Office Rooke had already learned about the Bonnybridge infections and had taken steps to find out more. When he came into work on the Tuesday morning he had tried to contact Bryceland at the North Lanarkshire Council offices using the emergency number for local authorities that was kept in a special green folder. He was unable to get through, and left his secretary to make more attempts but was unable to make contact by telephone all morning. So he then used the fax number in his emergency list and sent a message asking Bryceland to contact him urgently at the office. He received no response so he sent another fax at about 11.45 AM repeating his request for Bryceland to contact him as well as a letter indicating that as the outbreak appeared to have gone beyond North Lanarkshire's boundaries, he really should be kept informed about developments. Rooke's need to get a reply from Bryceland was overtaken by events when Ahmed's invitation to him came through at about 3 PM to attend a meeting with the outbreak control meeting at the Health Board Offices at 5 PM. Rooke attended. A list of places supplied by Barrs was drawn up to form the basis of an official food hazard warning. Because of a legal requirement that such a list should be complete—and it clearly was not going to be—it had to be marked confidential. Rooke faxed it to all Directors of Public Health, Heads of Environmental Health Departments and public health doctors the following afternoon.

From Tuesday onwards the Scottish Office took a deep interest in the outbreak. The wisdom of this was forcibly re-emphasized the next day, Wednesday 27 November. Three more patients died, two in the early hours and one at 3 PM. Two had attended the church lunch and had been admitted to Law Hospital. Alexander Gardiner had fallen ill a week before with bloody diarrhoea. The general practitioner thought he might have piles. He was chesty. The next day he had stomach pains and was vomiting and had become dehydrated. He was admitted to hospital where he developed fatal kidney failure. Mr Gardiner was particularly unlucky in view of his previous medical history because he was a member of that very select, small and fortunate minority of patients who had developed lung cancer and been successfully treated. His right lung had been removed in 1989. Jessie Rogerson was pretty healthy apart from a touch of back problems. She had also been at the church lunch. She fell ill the previous Tuesday with a sore stomach. By Thursday she had developed bloody diarrhoea. Her general practitioner, like Mr Gardiner's, thought that the blood might have come from piles. She had had them before. She became very weak but it was decided not to take her to hospital because the weather was very bad. By Saturday there were signs of improvement. The diarrhoea had stopped and she ate breakfast. But the improvement did not last and on Sunday her immediate admission to Law Hospital was arranged. But an ambulance was not available so she was taken by relatives in their car. Her kidneys failed and before her planned transfer to a renal unit in Glasgow could be done she died. Marian Muir lived in Cleland, a small town about three miles to the north of Wishaw. She was 70 and had a mild dementia and angina. She ate cooked meats from Barr's and had them on 15, 17 and 18 November. She felt unwell on the 18th, developed diarrhoea on the 19th and was admitted to hospital on the 21st. Her kidneys failed and her bowel suffered serious damage. She died at 1.15 AM on the 27th.

By Wednesday media interest in the outbreak was getting intense. They had picked up that the food hazard warning was confidential. There was beginning to be talk of 'cover ups'. The number of cases was still increasing rapidly and those responsible for outbreak control were not in a position to say that things would not go on getting worse. Even if they had known the names of all the individuals exposed to meat from Barr's and to meat contaminated with *E. coli* O157 from meat from Barr's—which they clearly did not—it would not have been possible to say with any degree of certainty how many of them had been infected, how many would fall ill, how long their incubation periods would be, and how many would develop complications.

Government ministers had begun to consider what action they should take. The option of being seen to be doing something new and important was rather limited because a lot was already being done. Civil servants in Edinburgh, like Stephen Rooke and his staff, were already deeply involved. Although there were compelling reasons why local control measures had to remain in the hands of local people, even so, those on the ground were already being helped by epidemiologists from the centrally funded Scottish Centre for Infection and Environmental Health in Glasgow. The *E. coli* O157 Reference Laboratory in Aberdeen—also centrally funded—was already working overtime to

Figure 1.2   *E. coli* O157 Reference Laboratory in Aberdeen (ground floor, far right behind tree) and refrigerated container holding meat from Barr's.

confirm and type the identity of bacteria isolated from patients' samples and was carrying out more sensitive non-routine diagnostic tests, and its sister public health bacteriology laboratory in Aberdeen was testing the swabs taken at Barr's, food samples and raw meats. It eventually received a massive lorry load of meat—5 tonnes—from Barr's under police escort. It was far too big to be stored in its walk-in cold room and a large 40-foot refrigerated container had to be hired (Figure 1.2).

So Michael Forsyth, the Secretary of State for Scotland, did what ministers in a similar position had done before time and time again for a hundred years and more. He set up an Inquiry. Early on Thursday afternoon I received a phone call from Rosalind Skinner, a senior medical officer at the Scottish Office. She asked whether I would chair an expert group to look into the outbreak. I agreed. Michael Forsyth announced the setting up of the group to Parliament in Westminster later in the day. Official wheels had also been set turning in other quarters. On Friday the Procurator Fiscal at Hamilton decided to involve the police in the investigation of the outbreak. Five environmental health officers began working out of Wishaw Police Station on the following Monday to gather evidence that could be used in possible prosecutions. Details were logged into the Home Office Local Major Enquiry System (HOLMES) computer database. The system was run by a Detective Inspector and four Police personnel. It generated 'actions' which were followed up by pairs of Officers. Their findings were recorded in the database along with documents gathered in the course of the investigation such as witness statements and epidemiological questionnaires. One aim was to identify common elements in apparently unrelated lines of inquiry. Particular attention was paid to getting information from those who had allegedly bought cooked meats and/or pastry products containing meat directly from Barr's shop on Saturday, 23 November 1996, and from Barr's staff, the next of kin of those who had died, the guests and organizers of the Wishaw Old Parish Church Lunch, and the Birthday Party at the Cascade Public House. The Police also took photographs for evidential purposes of Barr's premises and a video recording was made of a Church lunch re-enactment. More than 700 statements were taken and recorded in the HOLMES system. The incident room at Wishaw Police Station operated from 2 December 1996 until 7 February 1997.

On Thursday, 5 December the Crown Office announced that a Fatal Accident Inquiry (FAI) would be held into the deaths arising out of the outbreak.

Under the terms of the legislation governing an FAI, the Sudden Deaths and Fatal Accidents Inquiry Act 1976, it is compulsory in only two circumstances, after a death in employment and after a death in custody. But one can also be held where 'it appears ... to be expedient in the public interest ... on the grounds that it was sudden, suspicious or unexplained, or has occurred in circumstances such as to give rise to serious public concern'. There is no jury. The Sheriff is required to make a determination in respect of each death

> setting out the following circumstances of the death so far as they have been established to his satisfaction—
>
> (a) where and when the death and any accident resulting in the death took place;
> (b) the cause or causes of such death and any accident resulting in the death;
> (c) the reasonable precautions, if any, whereby the death and any accident resulting in the death might have been avoided;
> (d) the defects, if any, in any system of working which contributed to the death or any accident resulting in the death; and
> (e) any other facts which are relevant to the circumstances of the death.

The stage had also been set for Barr to be prosecuted. On 10 January 1997 he was charged with culpable and reckless conduct arising from the alleged supply of cooked meats in relation to the function at the Cascade Public House. In line with practice the FAI now had to wait until these criminal proceedings were over.

So within two weeks of the outbreak first declaring itself, its catastrophic effects had been perceived to be so great that they had induced action by a cabinet minister. They had caused the full panoply of the law to swing into action. Two inquiries had been ordered. But there was still uncertainty about the future course of the outbreak. Had the actions taken so far stopped the spread of the organism? It turned out that they had. All the subsequent investigations into how confirmed cases contracted their infections showed that the last date on which a case could have been infected by eating products sold from Barr's shop was Monday, 25 November. From an outlet supplied by Barr's where there could have been cross-contamination from his meat it was Sunday, 1 December—with only 11 cases in total falling into this category after the previous Wednesday. But an incubation period of as long as 14 days meant that it was inevitable that new cases would continue to appear for some time. The Cascade Public House party on Saturday, 23 November and the consumption of contaminated cold meats by residents of the Bankview Nursing Home on this and the following day had lit the fuses of potential time bombs. Other unpleasant facts to consider were that a delay of a few days usually occurred between the initial course of an acute *E. coli* O157 infection and the onset of serious complications, and that the organism was well known for its ability to spread from person to person, with the first victims in an outbreak often infecting others in a second wave of infection. So even if everything had been done perfectly and no more infections were being contracted from the original sources there was still a strong possibility of very bad news to come. Pessimists were not to be disappointed.

The attendees at the Church Lunch were the hardest hit. Mary Jackson died on 2 December at Monklands Hospital. She had developed both of the main complications of an *E. coli* O157 infection, the haemolytic uraemic syndrome, and thrombotic thrombocytopaenic purpura. James Henderson died on 4 December, also at Monklands. He had pre-existing heart disease and although a heart attack was the primary cause of death, *E. coli* O157 made a significant contribution to it. Herbert Swanston died on 10 December of acute renal failure. His pre-existing liver disease was a contributory factor. Josephine Foster also died of kidney complications, on 14 December. Sarah Cameron had developed both the haemolytic uraemic syndrome and thrombotic thrombocytopaenic purpura. She died on 18 December. Nearly all these patients had complex and difficult terminal illnesses. Four had plasma exchange, one had kidney dialysis, one died after a period on a life support machine, three were transferred from one hospital to another for more specialist treatment, and one was transferred twice for the same reason. Of the 74 elderly and 13 helpers at the lunch 45 were infected. In addition to the eight who died another nine were ill enough to be admitted to hospital.

The Cascade party goers began to report their illnesses on 29 November. Although one person became seriously ill with life-threatening bowel complications, nobody out of the 129 attendees died. Indeed 11 of the 25 who were infected had no symptoms at all. Most of the 129 were young. The majority were between 15 and 19, with only 12 (9 per cent) being 40 or older. Likewise investigations on Barr's staff showed that infection in younger people of working age were often free of symptoms. Twenty out of the 39 had laboratory evidence of infection and *E. coli* O157 was isolated from 10 of them, but only 3 had any symptoms. The importance of age in itself as a factor in determining the seriousness of an infection was dramatically illustrated by Joan Blackwood's fatal illness. She died in Stobhill Hospital, Glasgow, on 14 January 1997, from pneumonia and the complications of an *E. coli* O157 infection, kidney failure and intestinal haemorrhage. Although it was never established how she got infected there was no doubt that her illness started before the end of November. Although she was 77, until her illness she was very fit. She had a part-time job as a cleaner in an estate agents office, getting up at five and six in the morning and walking two miles to her work. Further confirmation of age as a very important determinant of outcome came from the events at the Bankview nursing home. Three residents died directly or indirectly from the effects of infection, Alexander Nicol aged 79 on 6 December, Mary Paisley aged 83 on 8 December, and Helen Fraser aged 94 on 17 December. All the elderly residents with positive stool cultures had symptoms whereas only 3 of the 10 culture positive but much younger staff members fell ill. Three more very elderly people who lived in the community also died, Christina Wright aged 87 on 8 December, Mary Smith aged 90 on 28 December, and Rachel O'Malley aged 86 on 24 January.

Scientifically the most secure way to summarize the outbreak is to consider the confirmed cases—people from whom *E. coli* O157 had been isolated. Two hundred and two were from Lanarkshire. One hundred and seven of them had purchased meats from Barr's shop, 29 had attended the Church Lunch, 26 had bought meat from outlets supplied directly or indirectly by Barr's, 13 had been infected person-to-person or by eating food contaminated at home, 12 had been at the Cascade party, and 10 worked at Barr's. In only six was a link with Barr's never established. Of the 73 confirmed Forth Valley cases 70 had eaten cooked meats from Scotmid shops in the Bonnybridge area. Of the three others, one person could not be traced and one could not remember enough detail about what they had eaten. One case was probably infected person-to-person.

Not only were all these cases linked because they had been infected by *E. coli* O157, they had all been infected by exactly the same bacterium. Just as DNA in body fluids found at the scene of a crime can be typed to prove the presence of the criminal so the DNA of *E. coli* O157 can be fingerprinted to prove or disprove the exact identity of the bacteria recovered from the victims of an outbreak. But this organism was also found in Barr's premises and in food prepared by him. This was remarkable. Investigating a food poisoning outbreak is not simple. Usually the best evidence—the contaminated food—has all been eaten. In many instances the presence of a harmful microbe is intermittent, not being present in all batches of food made during the period under investigation, and even when it is there it might only be present in small amounts—sufficient to infect a person but difficult to detect in the laboratory. But at Barr's the investigation garnered a rich evidential harvest. The outbreak strain was found on the vacuum packing machine in the factory and on the boiler used to cook the Church Hall Lunch. Eight raw beef samples and a cooked ham from the front shop were positive as were six roast beef specimens supplied by Barr's to other outlets and one bought by a customer over the counter at the shop. Corned beef—not a Barr product but from a butcher who dealt with Barr's—was positive. The final definitive links in the chain of transmission were the presence of the organism on roast pork from a Scotmid shop in Bonnybridge and in gravy from the Church Lunch. Massive amounts of it were found in one of the portions that had been given to a helper after being decanted from the bags of stew shortly after their delivery to the Hall.

One of the biggest achievements of modern medicine has been the control of infection disease such as smallpox and polio. Most of us now live long enough to be carried off by age-related diseases like cancer. But many people in rich countries still die of infection. It is still correct to call pneumonia 'the old man's friend'. However many of the people who die this way often do not catch infection from someone else, but are killed by their own bacteria. A good proportion of us carry organisms with disease-causing potential—pathogens—in our throats or on our skin. *Neisseria meningitidis*, which causes meningitis and septicaemia, *Streptococcus pneumoniae*, which causes pneumonia and meningitis, and *Staphylococcus aureus*, which causes boils, carbuncles, osteomyelitis, pneumonia, and a host of other unpleasant diseases are easy to find by swabbing the throats, skin, and noses of fit and healthy individuals. These are 'latent pathogens'. The psychologist James Reason has used this concept in his analysis of the causes of accidents.

'Latent conditions' or 'resident pathogens' are factors that are silently present for a long time in an organization but which eventually conspire with local circumstances and active failures to defeat all the defence mechanisms to cause a disaster. 'Active failures' are different. They are committed by front-line workers and usually have immediate effects. The identity of their perpetrators is often obvious and so the apportionment of blame is, seemingly, straightforward. Latent conditions are generated higher in the organization and arise from strategic and top-level decisions made by managers. They can also be created by the actions of regulators and governments. By influencing local factors that promote errors they make active failures more likely and they make their consequences worse by their effects on the defences of the organization.

For Barr's, analysis under the headings of latent conditions, active failures and a particular local circumstance is straightforward. The local event that started everything off was the entry into the premises of the outbreak strain of E. coli O157 on raw meat. It is probable that this came from an animal or animals slaughtered at the end of October or the beginning of November at the Wishaw Abattoir. Because of active failures, ready-to-eat foods left the premises contaminated with live organisms. Significant problems at Barr's were identified in the very early stages of the outbreak by environmental health officers. The most important ones encouraged the transfer of E. coli O157 from raw meat to cooked, ready-to-eat products. In the factory the single set of knives, the single vacuum packing machine, and the weighing scales were used for both raw and cooked meats. There was only one set of knives in the butcher's shop. Like the butcher's block there, it was also used for both the raw and the cooked. The shelves in the walk-in chill at the back of the butcher's shop were used in the same way. On his first interrogation of Barr's staff Jeffrey Tonner found that hand washing was not done when cooked meats were handled immediately after raw. So it was not surprising to find that the wash hand-basin in the factory had no soap, the one in the bakery had neither soap nor drying facilities and that throughout the premises they all had domestic bib-type taps which ensure the recontamination of hands when the water is turned off. Tonner had also uncovered the washing-up liquid story. It was described with vigour by Sheriff Cox in his FAI determination:

> At the time of the outbreak Barrs did use the same surface for the manufacture of sausages and for resting joints which had been cooked. At the time of the outbreak these surfaces were not being cleaned and then a bactericide applied. What was being used was a biodegradable washing up liquid for cleaning work surfaces. The description 'biodegradable' in the eyes of Barrs' senior staff was synonymous with

'bactericidal'. The liquid in use was green in colour. There is no doubt about that. Mr Barr thought that about five years before the outbreak he had changed his supplier on the recommendation of a former employee who said that he could get a cleaning agent with the same properties but at a more attractive price.... There is some confusion in the evidence of the staff about the colour of the liquid in use before the outbreak. It is clear from Mr McLean's evidence and Mr Tonner's evidence that what was shown to them was a green liquid and it was not a bactericide. This was confirmed by analysis.... I think it is probable that the cleaner Sandra Preston was confused when she says that the pink liquid was changed for a green liquid and in fact it was the other way about and that that the change was after the outbreak. But in any event she says she had nothing to do with cleaning the factory part of the premises. That was apparently the responsibility of the butchers themselves. They had no cleaning schedules. Sandra Preston had had at that time no training in food hygiene and received no written instructions on how to perform her tasks. She cleaned the butchers' knives 'with bacterial washing-up liquid'. But it was Bert Hepburn who told her what it was and it transpired he did not know what he was talking about.... Although Barrs' cleaning methods were not put to the test after the outbreak the evidence led indicates their deficiencies. A cleaning schedule ought to have been provided. The schedule should have been compiled specifically with the aim of preventing cross-contamination.

In this part of his determination the Sheriff is now moving from active failures to latent deficiencies. He went on:

The scale of Barr's turnover was such that he ought to have engaged a food hygiene consultant at a much earlier stage to advise him on how to manage safety within his premises. He should have ensured that his staff had basic training in food hygiene including cleaning. I have no doubt Mr John Barr liked a clean shop and maintained a clean shop. What he failed to do was to maintain a safe shop and the main ingredient of his failure was ignorance of the requirements which would produce that result.

It is fair to ascribe the abundance of latent conditions in the business to this factor. Its massive shortfalls in training and its inadequate equipment and infrastructure might have their genesis in management decisions to save money. But the business was doing well. So it is likely that a failure to appreciate the risks that it was running played a very important part as well. After all, Mr Barr had been running his business in the same way for many years without causing an outbreak. It is also likely that this false sense of security played a part in his decision to deceive the local Environmental Health Officers about the scale and nature of his business so that he could evade registration under the Meat Products (Hygiene) Regulations 1994. These required the approval of premises and laid down conditions about their construction and layout, the facilities within them and the general hygiene condition. They required the separation of raw and cooked products either by time or by space. All these things would have to be written down and backed up with records and laboratory testing. Implementation of them at Barr's would have cost money. Consultants would have to be hired. These Regulations were the UK's response to an EEC Directive, which said that they were not intended to apply to shops or premises where 'preparation and storage are performed solely for the purchase of supplying the consumer directly'. In the UK Regulations relaxation was given to retail shops selling to other retail shops including caterers and shops buying food intending to sell it as a take-away. In the original regulations the precise meaning of 'take-away' was not given and following requests for official advice a Guidance Note was produced. It said that

> Enforcement Officers need not verify the take-away market supplied by up to one tonne or 50% of premises' product (whichever is less) each week if that supply is on an 'immediate consumption' basis—it being assumed that such supply must automatically be to a take-away market.

In November 1995 Mr Barr was asked to fill in an application form for Registration, because the environmental health officer responsible for him at that time thought that he might need approval. It asked him to declare the weekly tonnage of finished meat products that he made each week. He said that it was one tonne. How much he was actually producing was never established. A combination of being economical with the truth and bad record keeping on Barr's part saw to that. At the time of the outbreak he had 29 full and 25 half joints of meat in stock—not less than 744 lb—and one of the premises he supplied, Devines, had 42 joints, a total there of at least 672 lb. One tonne is 2205 lb. So if this is all that he was producing only 788 lb were going elsewhere. But in addition to Devines he supplied another 55 shops, 14 of them in the Scotmid chain. So it is reasonably certain that the one tonne figure was selected because it was the maximum that allowed relaxation from the Regulations rather than an accurate measure of production. He was interviewed about the application and said that he supplied other shops. He did not say that he was supplying wholesalers with cold cooked meats who sold them on to distributors who in turn sold them to butchers. But he was—sufficient for him to need approval. He was later seen by another EHO who asked him where the majority of his meat products were sold and was told 'to the final consumer and that most of them were sold over the counter'.

The Sheriff at the FAI concluded that

> Mr Barr, whom I did not find to be a credible witness on this subject appeared to have little or no recollection of the events surrounding the submission of this application. Mr John Barr struck me as a person who knew his way around his trade, would be very aware of regulations which imposed extra obligations upon him. He would not be adverse to framing answers to the questions asked of him designed to achieve the result least burdensome to his business interests regardless of their accuracy.

Understanding how meats were contaminated after cooking at Barr's is easy. They were handled on contaminated surfaces with contaminated hands and were cut with contaminated knives. But the events at the Church Lunch were different. Barr's staff had no direct contact with the steak after it had been cooked in its plastic bags. Church hall helpers cut them open and did all the rest. The steak was heated a second time. It came out of the oven bubbling with heat. Could the *E. coli* O157 have survived this as well as the cooking at Barr's? There were major deficiencies in the latter. Some of the heating elements of the boiler were not working. The temperature reached during the cooking process was not checked. There was no cooking protocol. The meat was not cooled afterwards. Although one attendee at the lunch thought the meat was tough, I have little doubt that the process it went through at Barr's was hot enough and long enough to kill all the *E. coli* O157 that could have contaminated it. The organism is not heat resistant. Boiling kills it stone dead in a few minutes. But one of the few indisputable facts about the Church Lunch is that the outbreak strain was present in large numbers in the gravy decanted from the steak after its overnight storage in the Church Hall conference room. From the microbiological point of view the likeliest explanation of this is that when the steak was pouring out of the cuts made in the bags the previous day, it picked up some *E. coli* O157 organisms that had contaminated the outside of the bags when they were being handled at Barr's after their removal from the boiler. A very few would do, because the organisms now found themselves in an ideal environment for growth. It was rich in nutrients and still warm. Standing overnight gave them ample opportunity to multiply with abandon. Under optimal conditions *E. coli* can divide every 20 min. Even with the pessimistic assumption that growth would be considerably slower—because the stew would cool through the night—and that the organisms only divided hourly, in the 20 h between the opening of the bags and the reheating of the stew they would have increased in numbers more than a million-fold. But could such large numbers of organisms survive the second round of heating in the Church Hall oven?

Definitely not. Unlike Mr Barr's boiler it was in good order. From all accounts the kind of temperature reached was enough to kill millions of bacteria in a few minutes. The steak was in the oven for well over two hours. The pastry was not eaten by two victims—including one who died. So it was unlikely to be the source of infection. How was the infection transmitted? The most plausible microbiological explanation is that microbe-rich decanted gravy—cold because it was at room temperature—was used to cool the hot stew. This seems to be the only way that the organisms could by-pass the lethal heating steps in the boiler and the oven. But when questioned under oath at the FAI the minister's wife said that this did not happen. So there is a mystery. The Sheriff concluded that some bacteria must have survived the cooking. I think it extremely improbable. But whatever happened, there is no doubt that the organism came into the Church Hall from Barr's. Latent conditions at the butchers helped it along—despite the existence of food laws designed to prevent their development and punish the offender. What had gone wrong? Deficiencies were not restricted to Barr's. The food safety systems set up to protect the public from unsafe enterprises like Barr's had failed.

Chapter 2

Why Disasters Happen

Governments have a deep interest in food safety. It is driven by public pressure. Politicians know from bitter experience that when they are in power and something goes wrong—like a major food poisoning outbreak—they will be held responsible. So they make food laws and create structures to police them. In Britain their enforcement has been a traditional role for local authorities. In the case of John Barr's this was the North Lanarkshire District Council and its environmental health officers. We have already seen them in action during the outbreak, investigating it and trying to stop it. But the main function of such an organization is preventive. By inspecting food premises they are supposed to make sure that their products are safe to eat. They are charged with giving advice to achieve this and securing compliance with legal requirements by enforcement action. Clearly in North Lanarkshire something had gone wrong, because they had failed to prevent the outbreak. Was this due to bad law? Were there novel or particularly unusual circumstances at Barrs that it would be unreasonable to expect enforcers to detect or react to?

Or could it be explained by the way the enforcers went about their business?

Sometimes laws are not enforced because they are bad or stupid. The same reasons make other laws difficult to enforce, and some are just antique relics of the past. None of these things could be said of general food law in 1996. The main statute—the Food Safety Act—had only come into law six years before. It was modern and flexible. It was one of Margaret Thatcher's best legislative initiatives. The immediate event that caused her to take a particular interest in food safety was a proposal at the end of January 1989 by Kenneth Clarke, the Secretary of State for Health, to have an inquiry into the microbiological safety of food. His reasons were not scientific, but political. During the previous summer health ministers had been made aware of a rapid increase of food-poisoning cases caused by a new kind of chicken *Salmonella*. The laying hens were not sick but the bacterium lived in their reproductive system and got inside a small proportion of their eggs. Handling the problem was difficult because although positive eggs were rare, they were common enough to be a significant public health problem. The bacterium—like all other Salmonellas—was killed easily by heat. But mousses and tiramisu and lightly fried eggs were popular. Matters began to come to a head in late 1988 when the Chief Medical Officer for England and Wales instructed hospitals to stop using raw egg in favour of pasteurized products. Television food programmes took up the subject. The scene was now set for Edwina Currie to intervene (Figure 2.1). Feisty and photogenic, she had been Parliamentary Under-Secretary for Health since September 1986, with responsibilities for preventive health care, including drug and alcohol abuse, mental health, hospital services including those for the elderly and children, and women's health matters. She enjoyed media work, and had been repeatedly asked to comment publicly about eggs but held off until she thought that there was an overriding obligation to warn the public. She did this on Independent Television News in early December. She said 'most of the egg production in this country, sadly, is now infected with *Salmonella*'. All hell broke loose. Although she did not say 'don't eat eggs', egg consumption fell. In her own words, in evidence to the BSE Inquiry, 'both MAFF and the producers took a dim view of my action; they regarded *me* as the problem, and that if I were removed the issue would go away. I was informed that writs had been served and I resigned from office on 16 December. Subsequently I discovered that there were no writs.'

In the immediate aftermath of all this it is not surprising that Margaret Thatcher took rapid action in response to Kenneth Clarke's request for a food safety inquiry. She decided to set up a Cabinet

Figure 2.1   Edwina Currie enjoying her new career.

Committee. Called the Cabinet Ministerial Group on Food Safety (MISC 138) it had its first meeting on 7 February. Thatcher chaired it. It discussed its work plan, the proposed inquiry into the microbiological safety of food, and salmonella in eggs. Thatcher then decided it would be useful for a smaller group to discuss food safety issues. Unlike MISC 138 it did not include the Minister of Agriculture. Salmonella in eggs was on the agenda, as it was on the next meeting of MISC 138. By early April it had been decided to proceed with a Food Bill. MISC 138 met again on 2 May when it was agreed to prepare and publish a White Paper describing the Bill. The Bill would implement the EC Inspection Directive, which had the aim of harmonizing the general principles of food inspection in the community. It went through all the usual stages and became law the following year. It was well drafted. It has worked well in practice. The Act itself was *not* the cause of problems in North Lanarkshire. *Implementation* of the law was another matter, to which we will return later.

When disaster strikes a mature enterprise that has been running for a long time without apparent problems, one of the first lines of inquiry to be pursued by investigators is to find out whether anything had been changed shortly before. Had new methods of working been introduced just before the disaster? It is well known that a critical point in the maintenance of safe working is the handover time when shifts change. Was this relevant? Or, like one of the chapter headings in James Reason's influential book 'Managing the Risks of Organisational Accidents', could it be an example of the situation where 'maintenance can seriously damage your system'? This lesson has been taught many times over since its first quantitative demonstration in 1943 by the operational research section of RAF Coastal Command, when it found that the rate of failure of anti-U boat aircraft was highest just after an inspection: "they were doing positive harm by disturbing a reasonably satisfactory state of affairs." Of the human activities universal in hazardous situations—control under normal conditions, control under emergency conditions, and maintenance activities—the last poses the greatest problem. Before considering Barr's it is instructive to consider examples that illustrate these things in detail.

Ill-thought through changes in working practices—even if well-meaning—can have catastrophic consequences. On about 20 May 1989 a 26-year-old man from the Fylde, the flat coastal part of Lancashire dominated by Blackpool Tower, was involved in a fight. He was left with bruising to his head, a headache, and dizziness for 24 h. On 3 June he developed double vision. His right arm had become weak, and he couldn't move his tongue to the right. Food and drink spilt from the right side of his mouth, and he couldn't swallow properly. He was admitted to hospital. After preliminary tests, the next day he was transferred to a specialist unit in Preston. High on the list of several possible diagnoses was that the earlier trauma to his head had damaged some blood vessels in his brain. Two hours later he stopped breathing. He was put on a life-support machine. By the next day all four limbs had become paralysed. He stayed in hospital for 43 days and needed help with breathing for 5 weeks. Eventually he made a very good recovery. But none of his symptoms had anything to do with his fight. He had had botulism. Like 25 of 27 other sufferers from this disease at the same time, nearly all from the north west of England, he had eaten some hazelnut yoghurt. His mother had some from the same carton as well—he had not finished it because of its unpleasant taste. She fell ill the day after her son.

The most seriously affected child in the outbreak was a 3-year-old girl. When she arrived at the specialist childrens hospital in Manchester 24 h after the onset of symptoms she was confused, needed help with her breathing and couldn't speak or swallow. Her eye muscles and her intestines were completely paralysed. Although many other patients in the outbreak were seriously ill, only one died, of pneumonia caused by stomach contents getting into her lungs.

All these patients had been poisoned by one of the most powerful toxins known. When eaten it passes into the body and homes onto a part of the junction between nerves and muscles and sticks to it irreversibly, stopping nerve impulses being translated into muscle movement. Its potency and long-lasting effect explains why the injection of minute amounts under the beauty trade name 'Botox' removes wrinkles for months on end. The toxin is produced by a bacterium, *Clostridium botulinum*, which is common in the environment. It survives adverse conditions by producing spores. These are tough. Boiling does not kill them. The canning industry is obsessed with them. It knows that ordinary cooking is not guaranteed to kill them, so it either cooks at high temperatures for a long time or uses recipes which make conditions inside the can unfriendly to bacterial growth. This stops them germinating and producing toxin while the can is on the shelf in the larder.

The hazelnut yoghurt turned into a killer because the recipe had been changed. The canned hazelnut conserve that it incorporated was originally safe. It was made from a mixture of pre-roasted hazelnuts, water, starch, and other ingredients which were heated in a half ton capacity vat to a temperature of 90 °C for 10 min. The mixture was then pumped into metal cans which were sealed and put into a retort of boiling water for a minimum of 20 min. None of these temperature time combinations were hot enough and long enough to kill all the botulism spores that might be there. But sugar was added to the conserve. This stopped any spores from germinating successfully. To meet customer demand for 'diet' products, however, a batch of 76 cans was made without sugar; the artificial sweetener aspartame was used instead. About half of them were delivered to the manufacturers of the yoghurt eaten by the patients. The aspartame-sweetened batch was made in July 1988 and stored at room temperature until its incorporation into yoghurt on 19 May 1989. During this time botulism spores germinated and produced toxin. Investigations after the outbreak showed that each yoghurt carton contained between 1750 and 3750 mouse lethal doses.

Had Mr Barr fallen into this kind of trap? Had he changed any of his recipes? Had cooking protocols been altered? In his evidence to the FAI he made it clear that nothing out of the ordinary had happened just before the outbreak. His business ran itself. It had been doing what it did in the way it did without change for a long time. Because of its many deficiencies it would have been a lot better if there had been changes. But there were none—and so the outbreak could not be explained this way.

What about maintenance? In late October, November, and early December 1937 there were 341 cases of typhoid fever in Croydon, a large town just south of London. Forty-three people died. Between a fifth and a sixth of the population were supplied by water from the Addington well. Repair and upgrading work had been going on there for quite a long time. In January a workman had died. Work continued but the regular waterworks staff were replaced with sewerman volunteers.

To keep the water level down it was necessary to use two pumps with a combined output of 156,000 gallons per hour. Rapid pressure filters and chlorination apparatus to clean the water and kill any bacteria (chlorine is particularly effective at killing typhoid bacilli) had been installed. They were linked so it was not possible to operate the chlorinator and the filters separately. But the maximum filter capacity was only 80,000 gallons per hour so it was not possible to filter the water when work was going on down the well. A programme of work to repair the pumping plant was started on 24 September. Until 15 October the water was pumped to waste while work was going on. But on the following day the arrangements were changed. Pumping of the unfiltered and unchlorinated water into the public supply began. The scene was now set for disaster. Neither the Medical Officer of Health or the Borough Engineer were aware of this change. In any case, despite being in overall charge the Engineer was only dimly aware that filtration might be affected by the work, and when he was planning it with the assistant water engineer and the Waterworks mechanical engineering foreman chlorination was never mentioned. Why this important discussion did not happen was never established, but was clear that the assistant underestimated the value of chlorination, and that the Engineer did not know that cutting out filtration automatically cut out chlorination. So until 3 November those inhabitants of Croydon that were supplied from the well received raw, untreated, water (Figure 2.2). The well had been sunk in 1885 and was 205 ft deep and 10 ft in diameter. Workmen went up and down it in a skip. Instructions had been given that they must not urinate or defaecate in the well, but unknown to the engineer a bucket down the well was used for the former. This was not all. Unknown to himself or anyone else one of the workmen was a typhoid carrier, of the kind that excreted the organism in faeces rather than urine. He had been infected during the First World War. After incubation periods had been allowed for, plotting the timing of onset of the typhoid cases showed an excellent fit with his presence in the well.

A Public Inquiry followed the outbreak. It did not find any eye witnesses willing or able to confirm that the workman had fouled the well with his faeces. But it concluded that 'there were ample opportunities for the deposit of faeces and that there could have been no desire to multiply the daily journeys up and down in the skip'.

The Croydon Public Inquiry concluded that 'there was both misunderstanding and lack of communication between the responsible officers of the Corporation in connection with the work'. This is a recurrent theme in the incubation periods of disasters. A much more recent one illustrates this dramatically and showed how maintenance

Figure 2.2   The Addington well being inspected by experts.

activities intended to improve safety had exactly the opposite effect because of communication failures. Though having nothing to do with food or water, the general lessons for food safety that can be drawn from the Piper Alpha disaster are powerful and important, not just because the circumstances before and during the tragedy but because of the safety improvements that it brought about.

North Sea oil platforms are complicated places. Their function is not just to drill for oil and send it onshore through a main oil line (MOL) but to separate the oil from the gas and water that also comes up the wells. The gas is dried and then further separated into methane, which is sent onshore through its own pipeline, and into condensate—more familiarly known as liquid petroleum gas—with propane as a major component. Condensate injection pumps compress it to a pressure of 1100 pounds per square inch (psi) and pass it through a meter into the MOL for reseparation onshore. All this describes what used to happen on the Piper Alpha platform (Figure 2.3). It was situated 110 miles north-east of Aberdeen and stood in water 474 ft deep. The production deck was 84 ft above the sea. It consisted of four modules. In 'A' were the wellheads, or 'Christmas trees'. Oil, gas, and water were separated in 'B', and the gas compressors, including the condensate

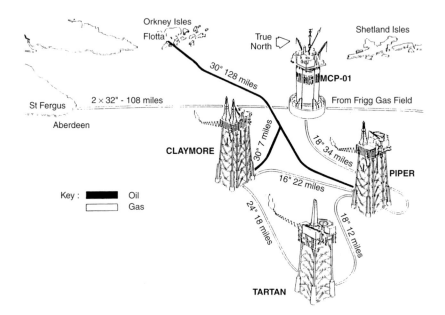

Figure 2.3   The location and pipeline connections of Piper Alpha.

injection pumps, were in 'C'. The power generators for the platform were in the lower part of 'D'; the Control Room was in the upper part.

On 6 July 1988 maintenance work was in progress. The 300 pressure safety valves on Piper Alpha were recertificated every 18 months, and a routine programme was just coming to an end. The last valve needing examination was PSV504, on one of the two condensate injection pumps. It had been removed and replaced with a blind flange—a flat plate held on with nuts and bolts—to cover the hole and stop the ingress of dirt. Blind flanges could be fitted in three ways, very tightly by flogging with a special spanner and a hammer, tightly by a combination spanner or loosely with the fingers. Disaster occurred because the last method was used. When the pump was turned on the flange leaked, condensate spewed out, and it exploded. But this should not have happened. The pump should not have been turned on because work on it was not finished. Things had moved on substantially since Croydon. A sophisticated permit-to-work system operated on the platform. There were maintenance diaries and operators logs, and formal shift handover procedures. But despite all these things, when the other pump tripped, PSV504's pump was turned on just before 10 PM by a production operator. Very soon afterwards, just after 10 PM, a range of unusual noises were heard by men in the Mechanical Workshop in 'D' module—'high pitched screaming', 'a banshee noise like someone strangling a woman', 'a high pitched grinding like metal to metal grinding together'. People in the Wendy Hut—the divers hut suspended at the 68 ft level—heard a loud hissing noise. Then there was a 'whoomph'. Between 30 to 80 kg of condensate had exploded. All this had happened because the permit-to-work system had failed. Information about what was being done had not been passed on at a shift changeover.

The explosion devastated the Control Room. The Control Room operator was thrown 15 ft by the blast. The other occupant, Alex Clark, the maintenance lead hand, was blown 6-8 ft against the oil well status board and hit with force on the shoulder and neck by the computer terminal. All the fire fighting systems on the platform were disabled. The main diesel fire pumps had been switched earlier from automatic to manual mode because diving was in progress and there was a risk of divers being sucked into the water inlets. Nobody could reach the pumps to turn them on because of the flames. Even if they had, many of the nozzles of the water deluge system in C Module were plugged with scale and did not work. Oil pipes had been ruptured by flying debris and fires started. Piper was linked by one oil and three gas pipelines to three other platforms, Tartan 12 miles away, Claymore 22 miles away, and MCP-01 34 miles away, and to the Flotta oil terminal 128 miles away in Orkney. A pool of fire developed under part of the Tartan riser—the pipe that came up into Piper from the seabed—and it ruptured at 10.20 PM. The explosion engulfed the platform in a massive fireball. The Tartan riser was at the same end of the platform as the East Replacement Quarters (ERQ), the main accommodation module and the place where many men had sought refuge. It was now at the centre of the flames. The pipeline contained gas at 1740 psi; in the first minute after its rupture 1.26 million cubic ft escaped. At 10.50 PM there was another explosion when the MCP-01 gas line ruptured—killing two crew on a rescue vessel—and the platform started to disintegrate. The biggest explosion of all was at 11.18 PM, when the Claymore riser ruptured. Not long after the ERQ fell into the sea (Figures 2.4 and 2.5). Of the 226 persons on board Piper, 165 died. One hundred and thirty-five bodies were recovered from the sea, the seabed, and from the ERQ after it had been lifted from the seabed. Post mortems showed that the principal causes of death were drowning (11), injuries including burns (11), and the inhalation of smoke and gas (109, 79 in the ERQ).

The differences between Piper Alpha and John Barrs are big and obvious. Burning hydrocarbons and *E. coli* O157 kill people in very different ways and with different timescales. Virtually all those lost on Piper Alpha had died within 90 min of the first explosion. The last person to

Figure 2.4   Piper Alpha platform, viewed from the south-east. The ERQ is at the near right corner under the helideck.

Figure 2.5   Piper Alpha platform after the explosion at 11.18 PM.

have a fatal illness caused by *E. coli* O157 from Barr's died two months after the start of the outbreak. The customers of Piper's operator, Occidental, suffered no physical harm from the disaster, which only affected those working on the platform. In Barr's it was the exact opposite. In Barr's, unlike Piper, there was no formal programme of routine maintenance. Some of the heating elements of the boiler used to cook the steak for the Church Hall were defective. Nothing was being done about it. Temperature probes were not being recalibrated—because there were none in the first place. It would have been far better if attention had been paid to these things. But errors in the *execution* of maintenance work at Wishaw were *not* the cause of the *E. coli* O157 outbreak.

Nevertheless, there are important parallels between Piper, Barrs, and Croydon. Bad communication was there in all three. The Croydon Medical Officer of Health did not know that the Addington well water was not being chlorinated or filtered. Neither did the Borough Engineer. Mr Barr 'blanched' when he realized that the 'bactericide' used by his staff was nothing of the kind. Alex Clark, the Piper maintenance leadhand, who had been smashed against one of the control room panels in the initial explosion and eventually escaped by climbing down a knotted rope to the 20 ft level and jumping into the sea, said when giving evidence to the Public Inquiry about the permit-to-work and handover systems 'It was a surprise when you found out some things that were going on.'

Organizations have learned through bitter experience, like Piper Alpha, that special attention has to be paid to shift handovers if operations in the face of ever-present hazards are to be conducted safely. The R101 airship disaster provides another dramatic example. Designed as the culmination of a programme started by a Cabinet Committee set up by Ramsay MacDonald in 1924, she was built to serve Britain's Imperial needs. When finished in late 1929 she was 732 ft long. It transpired that her lift would be insufficient for her to carry enough fuel to reach India and so in the summer of 1930 she was cut in two and a new bay and gasbag was inserted increasing her length to 777 ft and her hydrogen capacity to 5,508,800 cubic ft—truly a large inventory. For political reasons—the Secretary of State for Air wished to travel in her to India and return to England in time to attend the Imperial Congress in October, in his own words 'I must insist on the programme for the Indian flight being adhered to, as I have made my plans accordingly'—these alterations and the subsequent airworthiness trials were rushed. They had not been completed properly when it was decided to leave, despite concerns about gas leaks, on the evening of 4 October. The Secretary of State was accompanied in the airship by his valet as well as the Director of Civil Aviation, the

Figure 2.6   French policemen examine the wreck of the R101.

Director of Airship Development and her designer. In bad weather she headed from her base at Cardington in Bedfordshire over the Channel and across France. Seven-and-a-half hours later, at 2 AM, the watch changed, and a new height coxswain took over. Five minutes later she plunged into the ground and within seconds was engulfed in flames from end to end (Figure 2.6). Of the 54 crew and passengers only five engineers and an electrician from the rear of the ship survived. At the subsequent public inquiry the President of the Zeppelin Company, Dr Hugo Eckener, concluded that when the new coxswain took over, the ship was already nose heavy because of gas leaks and that he did not have time to 'feel his way into the condition of the ship' before it went into a dive so steep that nothing could be done to rectify it. All the coxswains perished in the accident but the dive was witnessed from the ground by a resident of Beavais, where the accident happened, Monsieur Rabouille. He worked in a button factory during the day, but happened to be out rabbit-snaring about 250 metres away from the crash.

A very important similarity between Croydon, Piper Alpha, the R101, and Wishaw is that by the time the first cases or the first explosion had occurred, whatever anybody did it was too late to make much difference to the final outcome. In Croydon the first local case of typhoid was notified on 27 October, the second on the 28th, and two more on the 30th. Fortuitously the carrier had stopped working in the well on the 26th, chlorination started on 1 November, and the well water was cut off from the public supply on 4 November. But the overwhelming majority of cases only began to fall ill later. They had already been infected before the first patients took to their beds. The lawyer in charge of the Public Inquiry commenting on the possibility that chlorination should have started on 30 October said that 'the dates of the various cases render it unlikely that any infection was caused by this short delay and if I had to come to a conclusion upon the matter I should say that it was not'.

The sort of accident that occurred on Piper Alpha—in essence a high-pressure fire—had been contemplated by those responsible for safety but their planning had assumed that any gas escaping in Module C could be quickly shut down and vented to platform flares so that it would all escape within 5 min. Cutting the gas supply off was essential because putting a fire out while it was still escaping could make things worse by allowing a gas cloud to grow which might then find another ignition source with catastrophic results. But even more important, as a memorandum of 18 March 1988 from Occidental's Facilities Engineering Manager to the Production and Pipeline Manager said, it was

> especially critical on Piper since we have no structural fireproofing...and all structural members are highly stressed. Structural integrity could be lost within 10–15 minutes if a fire was fed from a large pressurised hydrocarbon inventory.

This is what happened, of course. It also happened to the Twin Towers of the World Trade Centre on 11 September 2001. They collapsed when their structural integrity was destroyed by fires fed by the large 'hydrocarbon inventories' from the recently filled fuel tanks of two jet planes burning around the steel structures of the Towers from which the flimsy fireproofing had been blown away by the initial impacts.

In Wishaw the vast majority of cases—including all the guests at the Church Hall—were already incubating their *E. coli* O157 infections by the time the first sufferers fell ill. There was a brief window of opportunity when some other cases could have been prevented by shutting Mr Barr's premises earlier. He also broke the rules regarding the Cascade Bar. But as the Sheriff said in his FAI determination

> On the Sunday even if Mr Tonner had reacted straight away to Holloway's news by suggesting an immediate trip to Bonnybridge in his car with Holloway acting as navigator to the shops involved, the tragic events involving Bankview Nursing Home and Mrs Wright could not have been averted.

Even if Barr's had been closed within minutes of the first suspicions being aroused no lives would have been saved.

This pattern of events is common in food poisoning outbreaks. They are frequently caused by food served at a single meal. But many are due to organisms whose infections have incubation periods of several days. This period varies considerably from individual to individual. So even when all the victims are infected simultaneously, the onset of illness is dissociated from the event that caused it, and new cases go on appearing for quite a while. Even if the outbreak control team has confidently identified the contaminated food by interrogating the victims about what they ate, when the public sees people continuing to fall ill *after* reassurances given that all necessary steps have been taken and statements have been made 'that the outbreak is under control' they find official pronouncements hard to believe.

Over the last 150 years their expectations have been raised by the successes of science. Because of this they find it difficult to understand why old-fashioned diseases like diarrhoea cannot be nipped in the bud with speed. Many doctors hold the same view. Story-telling and myth-making about successful interventions during outbreaks have played an important role in building this view. Many medical textbooks retail accounts of how John Snow brought an outbreak of cholera to an end in the St James parish of London—better known now as Soho—in 1854 by causing the vestrymen of St James to remove the handle of the Broad Street pump, the source of the contaminated water causing the outbreak. It is a dramatic and satisfying story. It pays tribute to the effectiveness of the 'shoeleather epidemiologist'. It is commemorated today in an upstairs room in the 'John Snow' public house in Broad (now Broadwick) Street close to the site of the pump as well as by the annual 'Pumphandle Lecture' given to the John Snow Society in London on a date as close as possible to its removal, the 8th of September. But it is a myth. The outbreak was already nearly over when the handle was removed. In Snow's own words

> the greatest number of attacks in any one day occurred on the 1st of September, immediately after the outbreak commenced. The following day the attacks fell from one hundred and forty three to one hundred and sixteen, and the day afterwards to fifty-four...the fresh attacks continued to become less numerous every day. On September the 8th— the day when the handle of the pump was removed there were twelve attacks; on the 9th; eleven, on the 10th; five; on the 11th five; on the 12th only one ... the attacks had so far diminished before the use of the water was stopped, that it is impossible to decide whether the well still contained the cholera poison in the active state, or whether, from some cause, the water had become free from it.

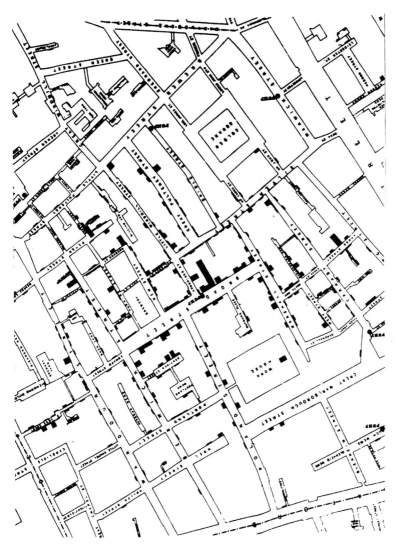

Figure 2.7   John Snow's map of Soho, September 1854. Fatal cases of cholera indicated by black squares. The Broad Street pump is at the centre of the map.

The prime responsibility for propagating this myth rests with the London physician Sir Benjamin Ward Richardson, friend and biographer of Snow. Comparing his and Snow's accounts of the meeting with the vestrymen is revealing. Snow said 'I had an interview with the Board of Guardians of St James parish, on the evening of Thursday, 7th September....In consequence of what I said, the handle of the pump was removed on the following day'. Richardson said

> While, then, the vestrymen were in solemn deliberation, they were called to consider a new suggestion. A stranger had asked, in modest speech, for a brief hearing. Dr Snow...was admitted...He advised the removal of the pump-handle as the grand prescription. The vestry was incredulous, but had the good sense to carry out the advice. The pump-handle was removed, *and the plague was stayed* (my emphasis).

Snow's influence among public health professionals is still great—not because of the pump-handle myth, but because his studies of the geographical distribution of cholera cases and their relationship to water supplies in London were—and still stand as—epidemiological masterpieces. He showed 'that the deaths either very much diminished, or ceased altogether, at every point where it becomes decidedly nearer to send to another pump than to the one in Broad Street'. He also found that the Poland Street Workhouse and the Broad Street Brewery were close to the pump but the inmates and brewery men hardly suffered at all (Figure 2.7); the workhouse had its own water supply and the brewer believed that his men didn't drink water at all—they were given free beer. Hampstead, distant from the pump, was free from cholera, but a percussion-cap maker's widow there died from it on 2 September. She liked the Broad street water and had had a bottle sent to her on 31 August. He felt that this was the most conclusive evidence incriminating the pump. Whether Snow was modest or not—and it must be remembered that heroes in scientific myths are often deemed to be so—he was a very successful pioneer anaesthetist. He chloroformed Queen Victoria during the delivery of her second and third children. She said 'the effect was soothing, quieting and delightful beyond measure'.

Rather than showing how outbreaks could be controlled, Snow's work emphasized how they could be prevented. His work caused him to conclude—many years before the causative organism had been discovered—that to prevent cholera 'care should be taken that the water employed for drinking and preparing food is not contaminated with the contents of cesspools, house-drains, or sewers, or, in the event that water free from suspicion cannot be obtained, it should be well boiled, and, if possible, also filtered'.

Chapter 3

Unlearned Lessons

It is evident that the particular causal circumstances which played key roles in the Croydon and Hazelnut yoghurt botulism outbreaks, the Piper Alpha disaster and the R101 disaster cannot be invoked to explain what went wrong at Barr's. Maintenance failures, shift handover problems, ill-thought-out recipe changes and lack of communication within the business were not specifically responsible. Neither were enforcement problems caused by defects in the law. So why did massive and prolonged cross-contamination, the main immediate cause of the Barr outbreak, occur in his premises in November 1996? The magnitude of the safety breakdown was gigantic. The first recorded sale of contaminated food from his shop to a microbiologically confirmed case occurred on 11 November, 11 days before the outbreak was recognized. Twelve days later *E. coli* O157 was still there in abundance when the Cascade Bar order was picked up by David Moon. During this time the organism had swept through the premises on a grand scale. It had got onto a range of meats including gammon, cooked ham, roast pork, and cooked turkey. It is certain that cross-contamination from beef was responsible because *E. coli* O157 is very rare in pigs and turkeys. In any case it is killed by cooking. It was also found on the top of the boiler and on the vacuum-packing machine. It is very likely that it was put there during the quick 'clean' that happened just before the environmental health officers descended on the premises early on the morning of Saturday 23 November—more evidence of cross contamination.

How all these cooked meats got contaminated with *E. coli* O157 is clear. It was being transferred to them from raw beef by unwashed hands, uncleaned knives, and contaminated working surfaces. But practices at Barr's had not changed for years. He had not caused any outbreaks before. So is cross-contamination such an unusual cause of food poisoning that its rarity excuses John Barr's ignorance of how to prevent it and the North Lanarkshire environmental health officers' failures to detect the consequent bad practices in his business? The question can be put another way. Even if such outbreaks had occurred from time to time, had knowledge of their details remained unpublicized? Were their circumstances only familiar to specialists like epidemiologists and microbiologists and academics working in ivory towers? Was it reasonable to excuse the failures in North Lanarkshire because it would be unreasonable to expect knowledge of cross-contamination as a risk factor to be widely held or understood? The answers to these questions is a resounding 'no'. Consider the classic Scottish case, the Aberdeen typhoid outbreak.

On 12 May 1964 an Aberdeen University student was admitted to the sick-bay of her hall of residence with a carbuncle. After a few days it became clear that her fever needed more investigation and so on the 16 May she was transferred to Aberdeen Royal Infirmary. On the 14 May her room-mate fell ill. She was admitted to the sick-bay on the 15th, and because of her unexplained fever was also transferred to the Royal Infirmary. By Tuesday the 19th a diagnosis of typhoid fever was being seriously entertained and both women were transferred to an infectious disease ward at the City Hospital, where the diagnosis was confirmed on the 20th. Two more definite diagnoses of typhoid fever were made on the same day and by the evening a total of 14 people with a confirmed or suspected diagnosis were in hospital or going through the admission process. Day by day for the next week admissions rose sharply. At the end of May 238 patients were in isolation. The outbreak then slowly waned with 294 being admitted in June and eight in July. By its end the total number of cases was 507. Early in the outbreak suspicion fell on a supermarket at the west end of Union Street in Aberdeen. Nearly all the patients (38 out of 41 admitted up to 23 May) remembered eating cold meat bought there, and three-quarters recalled eating sliced corned beef. But after about Tuesday, 26 May the pattern changed. More and more histories did not mention corned beef but only referred to other cold meats from the supermarket. Typing of the typhoid organism—*Salmonella typhi*—pointed to a bacterium of South American origin, confirming that the organism had come to Aberdeen in a can of corned beef from Argentina. Its contents were sold sliced between 6 and 9 of May, infecting the early cases. The same slicing machine was also used to cut other cold meats. In the process it contaminated them. The bacterium was then helped along because during shop hours some of the meats were put in the window in an uncooled display case. Warmed by sunshine this became an incubator. So the riposte to the joke that 'only in Aberdeen would you get 507 slices out of a can of corned beef' is that only 50 supermarket customers were infected from this source, while more than 400 were infected by eating other meats that contained the organism put there by cross-contamination.

Although no deaths during the outbreak could be directly attributed to typhoid, it attracted enormous publicity. In the main this was due to the actions of the Aberdeen Medical Officer of Health, Ian MacQueen. Because he feared that sufferers from the infection would spread it and cause a giant 'second wave' of infection he mounted a massive media campaign, holding daily—and sometimes twice daily—press conferences. He appeared on television every night. Local newspapers published wartime-like casualty figures with the names and addresses of the latest confirmed cases. He proclaimed the end of the outbreak by announcing, to use his own words, the 'All Clear'. MacQueen claimed that one of his aims was to 'allay panic'. But in this regard his campaign was a miserable failure. It had exactly the opposite effect. The headline 'Now a Beleaguered City' appeared in newspapers on 30 May. Rumours spread in the United States that there were so many dead that bodies had been piled on the beach waiting for burial. Offers of food parcels came in. Identifying oneself as an Aberdonian furth of the town engendered hostility. More than 25,000 visitors cancelled plans to spend their holidays in Aberdeen. There were leaders in *The Times*. In an attempt to achieve closure at the end of the outbreak a Royal event was arranged. The Queen visited Aberdeen on 28 June (Figure 3.1). But the consumption of corned beef in the United Kingdom still fell by more than half. Beef prices went up because consumers turned to other meat, which at the time was in short supply. Cattle raisers in countries specializing in exporting beef for canning—Paraguay, Kenya, and Tanganyika—suffered economic loss. With an impact of this magnitude it would be quite wrong to imagine that people in North Lanarkshire—even 30 years

Figure 3.1   Women wearing gloves. Her Majesty gives Aberdeen a clean bill of health.

on—wouldn't have heard of the outbreak and be familiar with its circumstances, at least in broad detail. Equally important, a Departmental Committee of Enquiry was held immediately after the outbreak that made recommendations about preventing cross-contamination. The relevance of those numbered (vi) and (x) to North Lanarkshire in 1996 is so obvious that they need no comment. They read:

> (vi) That there should be prepared in consultation with the trade a code of practice on the hygienic handling of cold cooked meats in retail establishments, and (x) only detergents and sterilisers whose bactericidal properties have been proved should be used in food premises; the responsibility for approving such materials should rest with the central health departments.

One line of action that followed Aberdeen was the initiation of a series of studies to give these recommendations scientific teeth. Richard Gilbert's 1968-70 papers were entitled 'The Hygiene of Slicing Machines, Carving Knives and Can-openers', 'Comparison of Materials Used for Cleaning Equipment in Retail Food Premises', and 'Cross-contamination by Cooked-Meat Slicing Machines and Cleaning Cloths'. The first made recommendations about the need for regular and effective cleaning of food-handling equipment. The second said that 'the purpose of this paper is to provide laboratory evidence of the obvious—that a contaminated slicing machine will easily cross-contaminate other products passed through it'. The third provided the evidence underpinning its conclusion that 'the effectiveness of any cleaning method depends not only on the cleaning agent used but also on its being used regularly and intelligently'. They were published in the *Journal of Hygiene*, one of the top bacteriological/public health journals. The work had been done by a leading food safety expert. They were clearly written. Their messages were simple. They clearly influenced food safety practice. But a quarter of a century later in North Lanarkshire they had been forgotten. Maybe they had never been heeded there—even when the memory of Aberdeen was still fresh.

Nobody died from cross-contamination in Aberdeen in 1964. It was very different 20 years later at Stanley Royd Hospital. Located near Wakefield, in Yorkshire, it opened in 1818 as the West Riding Pauper Lunatic Asylum. It was designed on the 'Panopticon' prison principle. In Jeremy Bentham's 1790s design the cells are arranged like the spokes of a wheel radiating from a central point. A solitary warder could watch all his charges without taking even a single step. At Wakefield the two intersections of the 'H' shaped asylum building were octagons containing central spiral staircases with iron crows nests from which staff could spy on other staff to prevent 'indolence and bad behaviour' (Figure 3.2). Almost from the start the hospital was too small for the demands that were being made of it. By 1822 the number of patients had exceeded the designed capacity of 150. The construction of extensions and additional buildings began in 1829. This process went on throughout the nineteenth century and into the twentieth. By 1936 it had 2678 inmates. Major changes occurred during the Second World War, and significant contraction started in the 1970s. But in 1984 the hospital still had 830 patients. Over a third were aged 75 or older and more than a quarter had been there for over 29 years.

The first case of diarrhoea to be diagnosed at the hospital was that of Thomas Umpleby, of Leathly. He was admitted a week after it opened, on Monday, 30 November 1818. On the next day he was noted to have a large and offensive stool. Looseness continued into December

Figure 3.2   A Stanley Royd 'crows nest'.

and on 9 January blood was seen in his stools. This continued. He lost strength and died on 9 February. In one of the many papers written on diarrhoea in lunatic asylums at the beginning of the twentieth century, H. S. Gettings, the asylum pathologist, reviewed diarrhoea at Wakefield. Writing in 1913, he concluded that during the 95 years since Umpleby's admission the hospital had never been free from it. Nothing much changed in the next 71 years either. In the decade before the outbreak food poisoning had struck twice, in 1974 involving nine patients, one of whom had died, and in 1979 when 33 patients were affected. This outbreak peaked on the August Bank Holiday weekend.

In 1984 up to five cases of diarrhoea on any one day was not considered to be unusual. One patient had it on the night of Saturday, 25 August. But there was a dramatic change for the worse the following day. An outbreak started which, in the words of the subsequent public inquiry report, was 'frightening in its scale and the rapidity of its onset'. The first patient's diarrhoea started at about 7 AM. By 9.15 AM 36 patients in eight wards had been affected. The numbers rose inexorably through the rest of the day—to 45 by 1.30 PM, 50 by 4.30 PM, 70 by 6.45 PM, and 94 by 9.15 PM. The first fatal case died at 11.35 PM. Sunday is not a good day to fall ill in a hospital because it will not be fully staffed. But at Stanley Royd the prospect of automatic relief coming the next day, Monday, was not there either. It was a Bank Holiday. During it the number of cases rose to 153. Hands-on medical care of these patients fell initially to Dr Peter Calveley, the duty junior medical officer. He had qualified the year before and had been in post at Stanley Royd for three weeks, on a general practitioner rotation. He was the only doctor in the hospital whose main job was to look after the physical health of the patients. Within half an hour after being called to the first case, at 8 AM on the Sunday morning, he had seen six or seven others in different wards. By 9 AM Ivy Ward was calling for nursing help as well because it had so many patients with diarrhoea and vomiting. The Duty Nursing Officer was told. By 10 AM she had set the wheels in motion for implementing the procedures laid down in the hospital Cross Infection Manual and had called a meeting of the Cross Infection Committee. Dr Fiona Davis was the other junior medical officer. Like Peter Calveley, she had been qualified for a year and had only been in post for three weeks. Her duties were to deal primarily with admissions, casualty calls, and the acute psychiatric ward. At 11 AM Calveley called on her for help. She agreed, and rescheduled her duties. They both attended the Cross Infection Committee meeting which began at 1.30 PM. It lasted about half an hour. During it Calveley had four or five calls from the wards. It also became clear that nurses were affected. Eight of them who had worked on the previous day's afternoon shift had been sent home with symptoms or reported in sick. Bacteriological sampling of food from the kitchen was started. The two doctors returned to the wards. The public inquiry report describes them as 'almost running from patient to patient, examining, prescribing, arranging for the monitoring of pulse, blood pressure, temperature and fluid balance, setting up drips and dealing with minor injuries caused by patients slipping and falling on wet or slippery floors'.

The second meeting of the Cross Infection Committee, now renamed the Infectious Diseases Committee, was held at 10 AM the following morning. Like the outbreak, it had grown. Its membership now comprised the Wakefield Health Authority District Medical Officer (in the Chair), the on-call Stanley Royd Senior Medical Officer (a psychiatrist), the Duty Nursing Officer, the acting Unit Administrator

(no senior administrator for the hospital was currently in post),
Drs Calveley and Davis, the acting Director of Nursing Services, the
Senior Nursing Officer (Occupational Health), a Supplies Officer, the
acting District Linen Services Manager, the District Pharmaceutical
Officer, and a local Consultant Pathologist. There was nobody from
either catering management or the kitchen. During the meeting the
local Environmental Health Department was contacted. An environmental health officer, Michael Greaves, arrived 40 min later. At the
end of the meeting it was agreed that examination of the kitchen and
what went on there were to be handed over 'lock, stock and barrel' to
the environmental health officers. The outbreak peaked the next day,
Tuesday, 28 August. Two hundred and forty patients were ill and
26 nurses and two domestic assistants had reported sick. Two of the
three wards in the hospital that had escaped infection in the first two
days of the outbreak, Gorse and Juniper, were now affected.
Conditions in the hospital were terrible. The public inquiry report
gives a graphic description:

> A number of members of the nursing and the domestic staff, both quali-
> fied and unqualified, gave evidence before us. It is unnecessary to report
> their evidence in this report, for conditions on a severely afflicted ward,
> where most of the patients were normally doubly incontinent, incapable
> of unassisted movement, and unable to follow even the simplest instruc-
> tions on personal hygiene, will be appreciated as defying description.
> We hope it will suffice to point out that many of the patients at Stanley
> Royd were confused and/or irrational, conditions which were exacer-
> bated by their suffering sometimes violent attacks of diarrhoea and
> vomiting, by the suffering of those about them, and the general turmoil
> and upset in what many perforce regarded as their home. Some refused
> to stay in bed and insisted on getting up, walking about and watching
> television. It was impossible to prevent patients from smoking, from
> sharing cigarettes, and from using whatever drinking vessels came to
> hand. Some patients physically resisted medication or attempts to make
> them use commodes or bed pans. Other, when moved to 'isolation'
> areas, made determined and persistent efforts to return to their own
> beds. To speak of barrier nursing against such a background is absurd.
> In addition, the moods and attitudes of the patients were not helped by
> the unusual isolation of the hospital, the restrictions on leaving the
> wards, the curtailment of activities, and the remarkable spell of hot
> weather which prevailed at the time.

On Tuesday the provisional diagnosis of *Salmonella* made on
Monday evening was confirmed. On Wednesday the number of
patients with symptoms began to fall. There were now only 185. But
53 new cases were diagnosed, and patients in the only ward so far unaffected, Almond, began to fall ill. Nevertheless, sighs of relief began to

be breathed. The outbreak seemed to be running out of steam. There were only 14 new cases on Thursday. But a patient who had been transferred the day before to Seacroft Hospital, which had an Infectious Diseases Unit, died. Three more died on Friday, one on Saturday, four on Sunday, one on Monday, two on Tuesday, two on Thursday, one on Friday, and two on Saturday. The last death occurred on Sunday, 9 September, when the last non-fatal cases in the outbreak—six—also fell ill. In total 355 of the 788 patients in the hospital had been ill and *Salmonella typhimurium* phage type 49 had been isolated from 218 of them. It had also been found in the stools of 81 hospital patients without symptoms. The deaths of 19 patients were due to, or contributed to by, infection with the organism. One hundred and six of the 980 staff had had symptoms, 80 with positive bacteriology of whom 51 had been ill. The outbreak was caused by contaminated roast beef that was served cold to the patients on the evening of Saturday, 25 August.

The precise chain of events that led to *Salmonella* getting onto the meat will never be known. The public inquiry heard much evidence about it and observed 'several dramatic and important conflicts' between the accounts it was given. They concluded that 'there were undoubted errors of recollection . . . in view of the lapse of time' but that there was also 'a strong element of self exculpation'. But there was overwhelming evidence that the beef served on Saturday afternoon was responsible.

The initial stages of the investigation were not straightforward. There is no dispute about the incubation period—the interval between infected and falling ill—of a *Salmonella* infection. It ranges between 6 and 72 h and is usually between 18 and 36 h. Neither was there any doubt about when patients developed their symptoms. Putting these times together narrowed suspicion to food served on Saturday. But finding out what people had eaten on that day was very difficult. Many special diets—fat free, minced, diabetic, and vegetarian—were being served. The nature of the patient population meant that it was often impossible to find out whether a particular person had actually eaten what they had been served. It was difficult to pin down what the staff had eaten because the kitchens had departed from scheduled menus, and staff were reluctant to admit that they ate meals originally destined for patients—a practice against the rules but not uncommon and because kitchen staff, feeling under attack, were reticent to divulge catering arrangements at the material time. But conclusive evidence pointing to the Saturday afternoon beef was found. A party of patients and staff had left for a seaside holiday in Lancashire at 10 AM on Saturday morning. None of them fell ill. With only one exception, none of the staff who fell ill in the first two days was on duty on Friday or Saturday when breakfast or lunch was available, but were on duty when the meal was served on Saturday afternoon. The exception handled the Saturday afternoon meal. Two staff members admitted eating the beef but had not consumed any other relevant foods at the hospital. They fell ill the next day.

The beef was delivered frozen to the Stanley Royd butchery department on 21 August. On the Thursday before the outbreak it was removed from the freezer to defrost. On Friday morning it was taken into the kitchen. It was unwrapped and the individual joints were cut into three or four parts. At about 11 AM it was put into the ovens and cooked for between three and three-and-a-half hours. No-one in particular was responsible for this and no strict timings had been laid down. After removal from the ovens the meat was put on a trolley and left in a store at room temperature until the following morning. Chickens were also defrosting in the kitchen on Friday morning. They were on a table used for general food preparation. Their juices drained onto the floor and into the gulley that crossed the kitchen. These were deep open drainage channels covered with metal grilles. They had a resident population of cockroaches which had been attacked from time to time by a firm of pest control contractors. Success, however, had been limited. Later, the cockroaches became attractive scapegoats. But they played no role in the outbreak. Being of the oriental kind they could not climb; some were tested for *Salmonella* but none were positive. Chicken was also being handled in the kitchen on Saturday.

Detailed typing of the *Salmonella* that caused the outbreak showed that it had not come from beef, but from chicken. The public inquiry concluded

> that there was ample opportunity for the beef to have been contaminated from a chicken source after it had been cooked. It is also obvious, that, particularly having regard to the hot weather at the time, it was kept in conditions which could not have been improved upon from the point of view of allowing bacteria to grow and multiply. We cannot help feeling that the most likely method of transmission of the contamination was by the hand or knife of someone employed in the kitchen but, in the absence of incontrovertible evidence, it would be quite wrong to speculate as to by whose hand or whose knife.... The wide spread of contamination, if it needed encouragement, would have been ensured by the use of the slicer.

These were the active failures at Stanley Royd.

But the latent conditions that promoted them were there in abundance. The kitchens were constructed in 1865. Their basic structures had remained unaltered for more than a hundred years, with high ceilings and walls extending beyond the reach of any cleaner. The grille-covered open drainage channels crossing the floor were difficult to clean because their sides were of aggregate. They stank. The big 'long' store was a tunnel with brick walls containing many holes. Hot water and steam pipes ran through it. When environmental health officers took temperatures on 12 September it was 20.5 °C outside; in the main kitchen it was 26 °C and in the long store it was 29 °C. There was a walk-in cold room but it was far too small for the demands put on it.

People had been trying to sort out this mess for years. In August 1978 Douglas Dale, the Wakefield Area Health Authority District Catering Manager wrote to Alan Kilshaw, acting District Administrator of the Health Authority (Western District). He proposed that the kitchen be reconstructed, enlarged, and re-equipped at a cost of £155,000. His letter finished: 'the main kitchen is a culinary disaster area which has had only minor cosmetic work/replacements over the years. It will become increasingly difficult to maintain any reasonable standards, particularly of hygiene, as the years pass.' Dale's letter initiated an amazing six-year saga of pass the parcel, paralysis by analysis, and strangulation by red tape. Dale's project was put in tenth place on the 1978 District capital scheme. By November it had transferred to the Regional Small Schemes Programme, ranked fourteenth, well below the cut-off point for implementation. The August 1979 outbreak stimulated a plan by the District Works Department to spend £35,000 on improving flooring, drainage, and ventilation. Nothing happened. Dale wrote to Kilshaw in November reiterating the need for a major capital scheme, now estimated at £200,000. In June 1980 the District submitted an upgrading plan as one of its major capital scheme bids to Area. It ranked fourteenth, again, giving it no hope of acceptance. In April 1981 Area included an upgrading proposal in its Capital Programme proposals to Region costing £150,000, with priority nine. In October Region told Area that the project now featured as a small building scheme in its planning list for progress by a planning team. Area asked for the project to be delegated to them, but it was decided that Region would do the initial feasibility study. The cost was now estimated, tentatively, at £300,000. In June 1982 preparation of the outline brief started. A major dispute now began between Area and Region. They could not agree about what the design capacity of the kitchen should be. Further complications arose in November when questions were raised about the long-term future of Stanley Royd and two other hospitals on its site, Pinderfields and Fieldhead. By January 1983 the estimated cost of the original scheme had risen to £419,000. The Regional Project Planning Team met in June to discuss the upgrade. Eight options were considered—the status quo, a £400,000 refurbishment, a piecemeal refurbishment, a new Stanley Royd kitchen, a new

Pinderfields/Stanley Royd kitchen, closure and provision from either Pinderfields or an outside contractor, food preparation at Pinderfields with finishing at Stanley Royd, or provision of a new cook-chill system. By October it seemed that Region would only go along with a minimal scheme, but nobody was sure what this meant. A member of the Regional Health Authority, the Revd Bruce Petfield, wrote presciently to Alan Pritchard at Area of 'the possibility of potential criticism which might arise if for instance there was an outbreak of food poisoning which could be traced back to inadequate kitchen facilities'. But District held back from spending big sums because the project was now a Regional one. By February 1984 the cost of the 'minimal' scheme was now put at £472,000. The last project team meeting before the outbreak was on 23 July. It was decided that the preferred option was to build a new Stanley Royd kitchen costing some £600,000. The public inquiry concluded that these planning activities were 'a remarkable example of what well intentioned individuals can fail to achieve unless someone is charged with the responsibility of ensuring that careful attention to detail does not lead to a complete cessation of all activity other than the production of paper'.

The cold storage deficiencies in the kitchen area were a very important contributory factor to the outbreak. It is no coincidence that it happened during a heat wave. After the cooked beef had been contaminated with *Salmonella* from the chicken on Friday it was stored for at least 24 h at temperatures very suitable for the growth of bacteria. Just like the steak and gravy at the Old Wishaw Parish Church Hall, it is probable that their numbers increased a hundred thousand—maybe a million-fold—on the meat before it was served to the patients. But even if this was necessary for the outbreak to occur, it was not sufficient. Contamination had to happen first. This was not caused by either the antiquity of the kitchen or its manifold structural deficiencies. Bad working practices were to blame. The public inquiry report described important and dramatic examples. There was a shortage of cleaning cloths. When they ran out, dirty ones were rinsed or left to soak in soapy water and then used again to wipe down surfaces or equipment. No cleaning staff as such were employed in the kitchens and cleaning duties formed part of the kitchen staff's bonus scheme. But there were no cleaning manuals or instructions as to who should do what and when. The metal-topped tables in the kitchen were washed with a high-pressure disinfectant jet, scrubbed, washed down with water, and then wiped down with the squeegees that were used on the floors. Kitchen supervisors knew about this but had not stopped it. One of them described it to the public inquiry as a 'dirty trick'. Why were bad working practices endemic in the kitchen? It was understaffed. It also suffered from very high sickness rates. In 1983/4 absences through sickness and holidays was more than 35 per cent. In consequence, cleaning schedules were in arrears. Supervisory grades spent most of their time cooking and not supervising. Food was prepared at times that suited the staff rather than when appropriate for its consumption. On occasions the preparation of meat for cottage pie had taken three days. After the outbreak an environmental health officer had said that meals should be prepared on the day of consumption. The catering manageress agreed. But the assistant head cook said, according to the public inquiry report, 'that paper schedules did not work in a kitchen and that the Union would not accept the extra pressure of work involved'. Bad industrial relations had also interfered with training. Food safety refresher courses had stopped after a strike in 1982 and were still in abeyance when the outbreak occurred. Management responsibility for the Stanley Royd kitchen fell to the Catering Manager and her assistant. But in the words of the public inquiry report

> their tasks . . . appear not to have involved a great deal of participation in kitchen affairs. They passed through, rather than spent time in, the kitchen. They . . . were unware of the habitual practices of alternating the staff menus at the weekends, and of consuming unofficial free meals in the kitchen after 5.00 PM.

It would clearly have been beneficial if the Catering Managers had read Erving Goffman's classic, *Asylums.* Describing life in an American mental hospital, he said that

> attendants expected to eat some hospital food even though this was forbidden, and those with kitchen jobs were known to 'liberate food'.

Goffman called these practices 'secondary adjustments', defining them as

> any habitual arrangement by which a member of an organization employs unauthorized means, or obtains unauthorized ends, or both, thus getting around the organization's assumptions as to what he should do and get and hence what he should be.

The inquiry censured the managers:

> we regret that the catering manager and the assistant catering manager did not roster themselves for duty on any Saturday or Sunday or at any time after 5.00 PM on a week day . . . we found their performance in monitoring standards of hygiene and catering practices to have been minimal.

It found management deficiencies near the top of the line as well:

> that monitoring was carried out by waiting for problems to come to the Region was bad enough, but to leave the solution to the District, uncritically, was wrong in role and wrong in performance . . . we also

find it quite incredible that, apart from expressing regret and sympathy about the outbreak, the Region did not discuss the problem, the causes, or the action being planned or taken, at any meeting thereafter.

It was clearly content with Stanley Royd's position as a low priority, unfashionable, and cheap hospital (in 1983 it ranked 58th out of 66 large mental hospitals in England and Wales for costs per patient at £25.46 per day against the national average of £40.16).

The public inquiry report said that 'the majority of its buildings, including the kitchen, were over 100 years old and justified the description of "Dickensian" in many respects.' This description was particularly apt. Dickens had a deep interest in lunacy reform. He knew about Stanley Royd. In the 28 November 1857 issue of his weekly magazine *Household Words* he quoted from Samuel Tuke's instructions for its building as the Wakefield Pauper Lunatic Asylum: 'the regulation of an Asylum', says this tract,

> should establish a system of espionage, terminating in the public. One servant and one officer should be so placed as to watch over another. All should be vigilantly observed by well selected and interested visitors; and these should be stimulated to attention, by the greatest facilities being afforded to persons who, from motives of rational, not idle curiosity, are desirous of inspecting such establishments.

The problem in 1984 at Stanley Royd was not that 'interested' visitors were discouraged. It was that they never came. There was no interest. The public inquiry singled out the Regional Nursing Officer for the Yorkshire Region, Janet Hutton, for particular criticism in this regard:

> We could understand when we were told that ... Miss Hutton had not visited Stanley Royd before the outbreak for she had then only been in the post for eight months. But we would have thought that an outbreak of this size rendered her support and advice as being appropriate and we found it almost incredible that she had not visited the hospital during the outbreak, or prior to her giving evidence before us some eight months later.

The widespread publicity given to the outbreak makes it improbable in the extreme that information about it failed to reach North Lanarkshire. Pictures of hearses leaving the hospital were broadcast on the television news. Photographs of the kitchen appeared in the national press. They had been taken by a reporter and photographer in white coat disguises. The public inquiry was critical of this subterfuge, but concluded that 'the reporter's opinion, that the kitchen premises appeared to be being given a thorough clean but were taking a long time to reach the required standards, was confirmation of the other evidence we obtained'. The rat that was found and killed in the kitchen on 11 October achieved near-iconic status, particularly when it was found to be infected with the outbreak strain of *Salmonella*. But the public inquiry concluded that 'the condition of this creature was more likely to have been a consequence than a cause of the outbreak'. It was it's last victim.

However culpable cross-contamination was at Wakefield, it would be reasonable to suppose that its lessons for Barr's might have been seriously diminished by Stanley Royd-specific factors like its occurrence in a long-stay hospital with many doubly incontinent patients. The same could not be said for the enormous 1989 *Salmonella typhimurium* Definitive Type 12 food poisoning outbreak in North Wales and Cheshire in July and early August. It affected 640 people, 330 in Wales, mostly in Clwyd, and 310 in England, mostly in or close to Chester but also in Manchester. Seventy-four were admitted to hospital and there were three deaths implicating the organism. Like Stanley Royd, it occurred during hot weather, there were staffing complications at one local authority caused by industrial action, and cross-contamination played a central role. But its relevance to North Lanarkshire was more obvious and more direct because the cross-contamination occurred in a butchers business and the outlets it supplied. The source of the outbreak was Wynne Williams (Flint) Ltd, a firm of butchers in North Wales. Like Barr's it produced cooked meats, sold direct to the public, had its own distribution network to other retail outlets, and supplied products to wholesale distributors. One was in Flint, North Wales; it sliced meats on two machines, vacuum packed them, and delivered them—sometimes in unrefrigerated vans—to premises in Chester that included supermarkets, shops, a nursery school, and a hotel, to a Caravan Shop in Talacre, Clwyd, and to retailers in Manchester. Most but not all of the meats went out under the distribution label. The other distributor was in Widnes. It supplied a market stall in Chester Market that was very popular with travellers using the bus station close by. Many of the victims had bought cooked meats from the market stall. The Aberdeen situation was repeated here—but in reverse—in that *Salmonella* was isolated from tinned corned beef cut on its slicer; it was being cross-contaminated from cooked meats from the Flint wholesaler that had previously been cut on the same machine. The outbreak strain of *Salmonella* was also isolated from cooked meats—pork and ham— from the freezer at the butchers shop and from cooked meats supplied by the Flint distributor to retail outlets in Manchester. In November 1989 the owner of the butchers shop was fined £8,800 with £1,435 costs for five breaches of the Food Hygiene Regulations 1970. They had been found at a routine inspection by environmental health officers on

3 July 1989 and included the storage of cooked meat beside raw meat with the potential for cross-contamination, lack of cleanliness, and lack of hand-washing facilities. Another successful prosecution was brought in July 1990 for the sale, or consigning for sale, of unfit meat. The owner was fined £2,000.

The parellels between this outbreak and Barr's are remarkably close. But despite being successfully prosecuted—a rare event in food poisoning cases—the name Wynne Williams did not achieve Central Scotland notoriety. The firm was lucky in that the organism contaminating its products was *Salmonella* and not the more lethal and virulent *E. coli* O157. Even so, people died in the outbreak. But although much publicity was generated, it was about something else. What protected the firm from being the main focus of attention—and obscured the message that contamination kills—was a public row of epic proportions between the two organizations involved in investigating the Welsh part of the outbreak. Clwyd Health Authority employed the doctors and the Delyn Borough Council was responsible for environmental health. The many failures of communication that occurred between them were not contingent on sloppy working practices or poor management—as at Croydon, Piper Alpha, and Stanley Royd—but were deliberate. They were evidence of the bad working relationships between the two organizations that got worse as the outbreak grew, and persisted long afterwards. A Welsh Office internal review the following year said that

> there are a large number of issues on which the accounts given by the
> Delyn DC and the Clwyd HA do not agree. It has in general not been
> possible to resolve these disagreements...despite (them) the main pat-
> terns of events emerges clearly enough, and in some cases the fact that
> there is disagreement serves to point up with particular clarity the extent
> to which the arrangements for dealing with food poisoning outbreaks
> were not working as they are supposed to work.

The review concluded that whether the disagreements had any effect on the size of the outbreak could only be speculated about. Nevertheless, it identified the reluctance of the Council to identify Wynne Williams as the source of the outbreak both during it and for a long time afterwards as a 'notable feature'—the Clwyd and Chester District Health Authorities, the Communicable Disease Surveillance Centre in London, the Welsh Office and the Department of Health all felt the evidence to be strong enough to justify public identification— and it said that this disagreement might have confused the public about what it should do about meat in refrigerators and freezers. In its final chapter 'Lessons to be Learned and Recommendations' the review diagnosed the dispute and prescribed remedies using a legal paradigm. Resolution would be on the basis of the given rules according to standards of due process. Possible conflicts arising from professional or organizational rivalries were not discussed. But anyone familiar with food law enforcement and the investigation and control of outbreaks knows that they are common. This kind of dispute is only one of the many difficult issues faced by food safety inspectors. It is one reason why their lot is not a happy one.

Chapter 4

The Inspectors Fail

What went wrong at Barr's was not due to bad food law. If good modern legislation was there to protect the public by setting hygiene standards in food businesses and by creating incentives for good practice by punishing those who sold products dangerous to health, why were the laws not being enforced? It was not because Barr's was being ignored. In the two years before the outbreak his premises had been visited by inspectors eight times, and checked in an informal way by one of his big customers, John Devine. Devine was in and out of Barr's three or four times a year; the last occasion two months before the outbreak. He noticed nothing untoward. He said of John Barr at the F.A.I. that he 'was one of the highest of hygiene persons that I know'. He trusted him because he knew that he had been a meat inspector and President of the Wishaw Abattoir. Another industry check had been done by John Skedzieleuski, Scotmid's Fresh Food Controller. He visited Barr's in August 1996. He did a 'visual' inspection, looking for practices that might lead to cross-contamination as well as checking on the appearance of the premises and the staff and how they were boning and slicing meat. It was a courtesy call as well, because Barr had just become 'Butcher of the Year'.

Both Devine and Skedzieleuski knew Barr personally. This might have influenced their opinions about the safety of the food he produced. Not unreasonably, so might his long-standing high reputation as a butcher. And he was a magistrate. But these factors should have been irrelevant when it came to inspections by the food law enforcers, the local environmental health officers. In the two years before the outbreak they visited Barr's seven times. In 1995 the premises were scheduled to be looked at in detail twice a year. On 21 April the inspector was Richard Proctor. A man in his early 20s and of few words, he had graduated B.Sc. with honours in Environmental Health in 1994 and started work as a qualified environmental health officer in December of that year. He stayed in the premises for about 90 min, checking their cleanliness and their general appearance, that a bactericide was being used, and whether there was separation between raw meat and cooked meat processing. On the 2 May he wrote to Barr's listing the non-compliances with the Food Safety Act and the Food Hygiene (Scotland) Regulations 1959/78 that he had found. In the meat preparation area the wash-hand basin needed soap, a nailbrush, and towels (preferably single use). The double doors should be routinely closed, the car tyre should be removed, and a freezer needed defrosting and repairs to its lids. In the front shop there was the need to conspicuously display a notice indicating the minimum meat content of products on sale as required by the Meat Products and Spreadable Fish Products (Scotland) Regulation 1984. There was another visit on 15 May to check whether these deficiencies had been remedied. In a letter to Barr's on 25 May Proctor said that he was satisfied that they had been, but the defective flooring and damaged extractor fan in the bakery area that he had noticed in April still needed attention. A visit on 12 June showed that the floor had been mended but the extractor fan was still broken. Yet another visit on 11 July revealed that all was now well.

In November the time had come round for a six-monthly inspection. It was done by George Jorgenson, a basic grade environmental health officer. He was familiar with the premises and had visited them before, in 1989 and 1991. He knew that the layout was poor and that because of the way the butchery and bakery parts were arranged there were a lot of staff movements between the raw and cooked meat areas. But he did not comment on this in his report during this visit. No food safety non-compliances were identified for rectification. On this visit he also discussed the possible need for Barr's to be approved under the Meat Products (Hygiene) Regulations 1994. As we have seen, nothing came of this; John Barr was reticent and monosyllabic and economical with information when it suited. He was sent a letter on 31 January 1996 telling him that his premises were exempt from the regulations. On 25 January 1996 Richard Proctor inspected the premises. He was accompanied by Jane McGahan. She had graduated in Environmental Health in 1994 and was still a trainee. She had taken a particular interest in hazard and risk analysis during her degree course, was regarded as the office expert on it, and on this visit she spent time discussing with John Barr the implications for him of the legal requirement for properties of food businesses to carry out risk assessments to identify steps critical to food safety. This requirement had come into force the previous September following the promulgation of new regulations. Two things happened after this inspection. First, a letter was sent listing non-compliances. The refrigerator temperatures needed monitoring. The mould growth on the walls of the walk-in-chill had to be removed and its recurrence prevented. Butchery freezers needed defrosting and lid repairs. The butchery wash-hand basin needed repairing, and soap, like all the other wash-hand basins in the premises, and the cleaning fluids and brushes in the food preparation areas needed to be put into a cupboard. A return visit on 7 March showed that all was well. Second, Richard Proctor prepared a risk assessment report using the scoring system laid down in Annex 1 of the Code of Practice No 9 of the Food Safety (General Food Hygiene) Regulations 1995. Points are awarded under this scheme for potential hazards, methods of processing, consumers at risk, compliance with health and safety, structural aspects, and confidence in management. He awarded a score of 90–5 points better than previously—a result that regraded Barr's from Category A, needing inspection every 6 months, to Category B, only requiring an annual inspection. An important factor in the regrading was the good score of five given in the management category because there was a reasonable degree of confidence in Barr's management and control systems. Because of the resulting reduction in frequency the 25 January inspection was the last full one before the outbreak.

Why did all these inspections fail to prevent the outbreak? The words of the inspectors themselves are instructive. Under cross-examination at the F.A.I. by the Q.C. for Lanarkshire Council, Richard Proctor's evidence went as follows:

> In the factory area we would be looking at the layout to see if it was structurally in satisfactory condition and satisfactorily clean.
>
> Now, please carry on. Just tell the Court what else you would be doing?—I would work my way up through the area. If there was staff there I would look at them as well. I would look at their personal

hygiene to see that they were clean, that they were wearing suitable overalls, that they weren't days old and caked in blood.

Would that have applied to the butcher's shop or not?—Yes.

So you would have looked at things such as their clothing in the butcher's shop?—Yes.

What about their hygiene in the butcher's shop, the personal hygiene of the staff? Is that something you would have looked at?—If there had been anything we had seen they were doing which was bad, it would have been commented on. We wouldn't stand over each member of staff to ensure they were doing everything . . .

Just pausing there for a moment, would you describe for the Court what the normal reaction of staff in shops is when the Environmental Health arrive?—I think they find us very intimidating. They feel as if we are there to catch them out, and they tend to be very careful in everything they are doing—they are very cautious, they want to make sure everything is done correctly. That isn't a natural working environment.

You don't think you see them in a natural working environment when you are there—No.

Because if they are using practices they know to be bad they are not likely to be using them in front of you?—No.

Is that the basic position?—Yes.

Graham Bryceland, Proctor's boss, went even further when he was cross-examined by the advocate for the Wishaw Old Parish Church and the Bankhead Nursing Home.

You have already indicated in your earlier evidence that an inspection won't throw up very much, according to your view, because people are alerted to the visit of the inspector?—No, it would not throw up with regard to bad practices on the part of the staff, because they are going to be on their guard when there is an Environmental Health Officer on the premises.

Can I take it that your considered view is than an inspection is useless in relation to identifying bad practices on the part of the staff?—Yes.

Does that then suggest that there needs to be not fewer visits but more visits, so that the staff become more relaxed in the presence of an inspector?—Yes.

Proctor and Bryceland's views cannot be regarded as those of dispassionate analysts. They did not find much favour as excuses with the sheriff at the F.A.I. Neither did the position adopted by their employer—when the Q.C. for North Lanarkshire Council asked Jane McGahan whether a 'food hygiene inspection is only a snapshot based on what you can see on that day', not surprisingly she said 'Yes'.

But even if it is accepted that an inspection is an imperfect tool because of the circumstances and time pressures under which it is done, the North Lanarkshire environmental health officers were compounding the problem because they were inspecting the wrong things. Their emphasis on visible structural defects—the 'walls and ceilings' approach as it is now pejoratively described—betrayed an underlying philosophy that in many ways had not moved on since the days of Florence Nightingale. She placed contagion on the same footing as witchcraft and superstitions. Foul air soiling the walls and floors was the thing to fight. The first canon of nursing for the patient was to KEEP THE AIR HE BREATHES AS PURE AS THE EXTERNAL AIR (her capitals). Shades of the extractor fan! Florence Nightingale died in 1910. But clearly her miasmatic views are still with us.

'The limitations of sampling, especially on the basis of "what catches the eye" within a relatively short visit . . . runs a plain risk of missing what lies deeper than a surface inspection and of failing to reach a true assessment . . .'. Lord Cullen's words at the end of the first volume of the Public Inquiry Report into the Piper Alpha disaster apply equally well to the pre-outbreak inspections at Wishaw. The similarities between the events that took place during the incubation periods on the North Sea oil platform and in the North Lanarkshire butchers are quite remarkable.

Training in food safety at Barr's was inadequate. Safety training on Piper Alpha was the same. After the disaster 26 survivors were asked whether they had received a safety induction. Six said 'never'. Two were not sure. Of the 18 remaining 4 said it had lasted 5–10 min. One of them had arrived at Piper two days before the disaster. His briefing went as follows:

> He asked if we had been on the Piper before. I said 'No.'
>
> He said 'Have you worked offshore before?', and I said 'Yes'. He said 'Well, you will know what the score is then.'
>
> That was much about what it was.

Unlike Barr's, Piper Alpha was purpose built. But it was old-fashioned and over time lots of modifications and additions had been made which pushed it in the Barr direction. It had suffered a good deal from corrosion in its early life and had had to be repaired many times.

But it was the pre-outbreak and pre-disaster inspections of both that showed the greatest similarities. Like Barr's, Piper Alpha was inspected at regular intervals. They were carried out by the Safety Directorate of the Petroleum Engineering Division of the Department of Energy. As with environmental health inspections of food premises their frequency was determined by a rating system. Points were given on the basis of the type of operation, the effectiveness of management to maintain acceptable standards, the complement on board, and a general view of all aspects of health, safety, and welfare, including training, maintenance, and emergency procedures and equipment. The rating scheme was designed to ensure that installations 'at greater risk' were inspected more often.

R. D. Jenkins had become an inspector in the Safety Directorate in March 1987. Like all inspectors, on recruitment he immediately became a 'Senior Inspector'. He made his first visit to Piper Alpha in June 1987. So just like Richard Proctor and Barr's he had been in post for four months when he made his first inspection there. The Piper Alpha Inquiry Reports described what happened.

> He said that he found the platform was well run...The methods of working were not necessarily committed to writing. Although they were often based on custom and habit they appeared to be satisfact-ory....Following his inspection (he) discussed a number of compara-tively trivial points with the Offshore Installation Manager which he put in writing on 12 June 1987...On 10 July Occidental replied...This response was regarded as satisfactory.

On 7 September 1987 F. Sutherland, a rigger working for a contractor, died on the platform. A motor was being lifted to replace a pump bear-ing and he had climbed onto a panel that formed part of a canopy over the pump to attach a shackle and slings. The panel shifted, causing him to fall and sustain fatal injuries. Jenkins went out to Piper the next day and submitted his report three weeks later. He concluded that night- and day-shift handovers were uncoordinated and that more things were being done than the permit-to-work allowed. Occidental were prose-cuted under the Health and Safety at Work Act and pleaded guilty. Jenkins's next, and last, visit to Piper was on 26 June 1988, 12 days before the disaster. It was to check that the necessary improvements had been made since the Sutherland fatality and to do a normal routine inspec-tion. He spent 10 hours on the platform, taking a 'comprehensive walk' round the drilling and production areas and the 68 ft. level. This took most of the afternoon. He checked some permit-to-work documents against work being done but did not concentrate on this area because it was not regarded as a key factor in the Sutherland fatality. In his report he said that 'with respect to the fatality, the following improvements in working practices were noted: (a) handovers between shifts have been tidied up, (b) Occidental are looking at more formal methods of under-taking jobs through the job task analysis scheme...' He concluded:

> There appears to be a new air of confidence in Occidental with appraisal drilling and well testing both on fixed platform and from a number of semi-submersibles round about. Lessons appear to have been learned

from the Sutherland fatal accident. A routine inspection in one year's time is appropriate.

The optimism of Jenkins' final report was explained by J. R. Petrie, the Department of Energy Director of Safety, along exactly the same lines as taken by North Lanarkshire Council to explain Proctor's optimism about Barr's. He described an inspection as

> essentially a sampling exercise. The inspector samples and audits the state of equipment and working and management procedures. He talks to personnel and seeks to obtain an over-all picture of how well the installation is being operated, maintained and managed. An inspector must exercise his professional judgement in determining the scope and depth of the inspection . . . on our inspections we report what actually catches our eye at the time of the inspection.

In his F.A.I. determination Sheriff Cox did not mince his words about inspections.

> The practices in Barrs according to Mr John Barr were long standing. They were flawed. If they were observed by successive inspectors then there is no note in the file to this effect. If they were not detected they ought to have been. It is accepted that a worthwhile inspection of premises takes time—longer than was devoted to Barr's premises by inspectors prior to the outbreak. But with the emergence of organisms as dangerous as *E. coli* O157 there is no room for compromise. Inspections have to be thorough. Their nature may change or may already have changed with the introduction of the emphasis on risk assessment. But the end result remains the same. It is the production of safe food.

Lord Cullen was equally critical about the inspection of Piper Alpha:

> Even after making allowances for the fact that the inspection in June 1988 proceeded on the basis of sampling it is clear to me that it was superficial to the point of being little use as a test of safety on the platform. It did not reveal any one of a number of clear-cut and readily ascertainable deficiencies. . . . Apart from any other considerations, the length of the visit at that time was manifestly inadequate having regard to the size of the installation, the activities then taking place and the recent fatality.

Sheriff Cox also said

> that the more junior members of the inspectorate should have been trained to identify the risks which different businesses present . . . their performance ought to have been monitored to ensure that the inspections which they carried out were effective.

Lord Cullen in fewer words said exactly the same: 'the inspectors were and are inadequately trained, guided and led.'

Chapter 5

Inspectorates have Limits

The identification of strikingly similar patterns of failure with the inspection regimes applied to enterprises as different as butchers and oil rigs indicates the strong likelihood of deep-seated problems rather than isolated local difficulties. This raises a serious question about the utility of traditional inspections as deliverers of what legislators intend. Does the track record of inspectorates across the board support this view? What does history tell us?

Local and central government inspectorates in Britain have many functions and complex histories. But two main classes dominate; efficiency and enforcement. Efficiency inspectorates have little in the way of formal legal power beyond a basic right to inspect. Their main role is not to enforce the law but to make sure that public funds are being used for the purpose and in the manner intended by government.

Enforcement inspectors have statutory powers. One of their main functions is to see that particular legislative requirements have been met, and many of the inspectorates can initiate action through the courts if they are not. Although bodies of this type cover a wide range of government activities, the majority of their inspectors are concerned with health and safety. Those who work for central government often cover highly specialized fields through small inspectorates—like those for nuclear installations or animal cruelty. Local authority inspectorates are nearly always large and multi-purpose. Environmental health officers form the largest group. They have a wide range of duties falling under the general heading of 'public health'. As well as inspecting, they have an investigatory role. We have seen the North Lanarkshire team at work doing this at Wishaw. Licensing is another function of some inspectorates, including environmental health. Only a few present-day inspectorates were founded before environmental health. Lunacy was the first, in 1828, and anatomy the second in 1832, followed by factories in 1833, railways in 1840, and mines and quarries in 1842. The diversity of remits, structure, powers, and patterns of development of the members of this small group is large. It is big enough to illustrate in abundance the general principles that determine why inspectorates succeed, and why they fail. Their diversity is in large part due to the very different reasons that caused them to be established. Their enabling legislation was nearly always driven by a combination of reforming activitists— often parliamentarians—and lurid, often horrible, events and revelations that helped to drive public opinion and quell the voices of opposition. All these early inspectorates produced significant improvements. But for all of them success has been incomplete. The different inspectorates have failed in different ways. Looking at them in detail is a good way to analyse the limitations of inspection as a regulatory tool.

Ever since its establishment in 1842 in response to concerns about mine safety driven by the frequent and regular occurrence of mine disasters, the Mines Inspectorate had focused on what went on underground. Because many of the issues were highly technical, the inspectorate was largely made up of mining engineers. Its aim was to ensure compliance with regulations underground, and to investigate and report on underground accidents. But its remit was too narrow. In terms of the loss of human life probably the most important—and certainly the most harrowing—example of a catastrophe in Britain that could have been averted if it had been broader was the Aberfan disaster. At 9.15 AM on Friday, 21 October 1966, a waste tip from Merthyr Vale Colliery perched above the south Wales mining village of Aberfan suddenly slid down the hill, destroying first a farm cottage and killing its occupants, and then engulfing 20 houses and the Pantglas Junior School, where the children had just started their lessons. One hundred and forty-four people died including 116 children (Figures 5.1 and 5.2).

The Mines and Quarries Act 1954 was the main piece of legislation in force at the time of Aberfan. It did not cover tips. After the disaster

Figure 5.1   Aberfan: the tip and the village, 21 October 1966.

Figure 5.2   Aberfan: the mass grave, 27 October 1966.

the Inspectorate pointed out in its annual report for 1966 that Aberfan was not reportable under the Act because no miners had died or had been injured. It was admitted at the public inquiry that the tip was looked at from time to time to protect the men working there. But the evidence clearly indicates that the inspection was a desultory one. Tip experts knowledgeable in soil mechanics were never consulted. Tips had fallen through the regulatory net. There can be little doubt that this was to a large degree a consequence of the very close relationship between the Inspectorate—mostly made up of former National Coal Board employees—the miners themselves, and the Coal Board. The Inspectorate hardly ever prosecuted. It had undergone regulatory capture. This was a particularly bad failure because the disaster was a foreseeable one. The tips were unstable because they had been built over a spring. This was marked on an Ordnance Survey map of 1919. A tip slide of similar size had occurred nearby in 1939. The Aberfan tips themselves had slid in 1944 and 1963. Before the disaster their shape clearly indicated instability. In memorable words the public inquiry report summarized why the Inspectorate had not paid attention to these things:

> we found that many witnesses, not excluding those who were intelligent and anxious to assist us, had been oblivious of what lay before their eyes. It did not enter their consciousness. They were like moles being asked about the habits of birds.

The Mines Inspectorate was set up because mines were dangerous places. At the beginning of the nineteenth century so were lunatic asylums. The driving force behind the 1828 Madhouse Act was long-standing pressure from enthusiastic reformers, helped by a series of scandalous revelations that were reported in newspapers such as *The Times*. The best-remembered is the case of James Norris, a patient at Bethlem Hospital (Bedlam) in London. For 10 years he had been secured without relief in a close-fitting iron cage. According to the report of the 1815 Select Committee on Madhouses he

> stated himself to be 55 years of age, and that he had been confined about 14 years; that in consequence of attempting to defend himself from what he conceived the improper treatment of his keeper, he was fastened by a long chain, which passing through the partition, enabled the keeper by going into the next cell, to draw him close to the wall at pleasure; that to prevent this, Norris muffled the chain with straw, so as to hinder its passing through the wall; that he afterwards was confined in the manner we saw him, namely a stout iron ring was rivetted around his neck, from which a short chain passed to a ring made to slide upwards downwards on an upright massive iron bar, more than six feet high, inserted into the wall. Round his body a strong iron bar about

Figure 5.3   Print of James Norris published in 1814.

two inches wide was riveted; on each side of the bar was a circular pro-
jection, which being fashioned to and enclosing each of his arms, pin-
ioned them close to his sides. This waist bar was secured by two similar
bars which, passing over his shoulders, were riveted to the waist both
before and behind. The iron ring round his neck was connected to the
bars on his shoulders, by a double link. From each of these bars another
short chain passed to the ring on the upright iron bar.

Prints were published of Norris in his cage (Figure 5.3). They made a
deep impression on the public. But it was only one case of cruelty
among many. Patients at Bedlam were medically assaulted as well.
They were bled, purged, and given emetics indiscriminately. On questioning by the Parliamentary Select Committee the apothecary blamed
the physician, the physician blamed the apothecary, and they both
blamed the surgeon.

Commissioners were appointed under the Madhouse Act. They
were all members or ex-members of Parliament or doctors. Their main
role was to inspect private madhouses in London. But the number of
asylums in the provinces increased substantially over the next few
years, and the 1845 Lunacy Act established Lunacy Commissioners as

professional inspectors for the whole of England. There were five unpaid laymen, and three medical and three legal Commissioners, each paid £1500 a year.

One of the most successful Commissioners was Samuel Gaskell. Brother-in-law of Mrs Elizabeth Gaskell, the novelist, in 1840 he had been appointed medical superintendent of Lancaster Asylum. Opened in 1816 in externally attractive buildings of local sandstone fronted with an impressive four-pillared Tuscan pediment, inside it was grim. Patients slept on straw and many went barefoot. 'Wet and dirty'— doubly incontinent—patients were chained to commodes. Its first superintendent had established a strong tradition of physical restraint, preferring manacles to strait-jackets. This policy was maintained after his dismissal on corruption charges in 1825. His only concession to patients had been the invention of wrist- and leg-locks that eliminated the clinking of chains. Within a year of Gaskell's arrival all forms of restraint had been removed. Patients slept in cribs and were put to work. Gaskell paid a lot of attention to carrying the staff of the asylum with him in his reform programme. Their resistance to change was taken in hand and overcome despite the extra work involved. 'Keepers' were renamed 'attendants' and police-like uniforms abolished. 'Potting' of wet and dirty patients through the night reduced foul beds by orders of magnitude. Particular attention was paid to the kitchens. All this good work brought Gaskell national recognition. The regime he established at Lancaster was praised by Charles Dickens after a visit to the asylum in 1857. In an article in *Household Words* he said 'An immense place, admirable offices, very good arrangements, very good attendants; altogether a remarkable place.' Gaskell was appointed a Lunacy Commissioner in 1849. The first asylum superintendent to become an inspector, he set new standards for rigour and diligence. An inspection was done jointly by a lawyer and a doctor. They stayed on the premises for days. Gaskell was notoriously thorough. In his inspections he looked at everything. One of his obituarists said although 'as a Commissioner he was highly esteemed... by the superintendents of the institutions of the insane... (they) were at times disposed to resent his very thorough and minute examinations of the institutions he inspected from floor to ceiling.' Institutional food was eaten. Privies were visited. All the patients were interviewed.

Legally the Commissioners had few powers regarding the running of asylums. They worked mainly through persuasion. Their reports were published. Sometimes their criticisms were publicly rebutted. When this happened they responded vigorously. An exchange between the Commissioner and the Committee of Visitors of Colney Hatch

Asylum near London about an inspection made in 1861 is revealing. Gaskell had been one of the inspectors. The hospital defended itself by implying that an *annual* inspection cannot be a fair basis on which detailed recommendations for action can rest—the argument that the inadequacy of sampling inherent in an inspection amounts to a major flaw in the process. In its fundamentals this is the same defence used by the North Lanarkshire Council and the Safety Directorate of the Department of Energy after Piper Alpha.

> The Committee . . . observe, with much regret, that the more they have, from feelings of courtesy, listened to the representations of the Commissioners in Lunacy, the more that body has attempted to encroach on the functions of the Committee of Visitors, and to assume a tone of dictation . . . But the Committee must remind the Commissioners that they are the body to which, by law, is entrusted the care and management of the Asylum . . . and that to them the Court of Quarter Session and Ratepayers of the County look for a just and economical administration, which shall be not unsuited to the class of life to which the unfortunate Patients under their care belong. The Committee cannot conclude their remarks without expressing their surprise that the Commissioners should consider themselves able, after a visit once in the year, to express themselves so positively as to what alterations are desirable. The Committee, who, during the whole of the year devote constant and anxious care to the management of the Asylum, know that many of the matters suggested by the Commissioners in Lunacy are quite impracticable, and . . . [they] do not appear sufficiently to bear in mind the fact that Colney Hatch Asylum is established for Pauper Lunatics only, and that many luxuries and appliances suggested by them are quite unsuited to that class of Patients.

The Commissioners replied:

> The Commissioners, observing the warning that is given them in your letter against encroaching upon the functions not properly theirs, and . . . the claim of the Visitors to have treated their suggestions with 'courtesy', regret to be under the necessity of reminding the Committee that the members of the Board, on whom the Legislature has imposed duties of visitation and inquiry applicable to all Asylums, have a title to graver consideration than mere 'courtesy' . . . Nor can the Commissioners regard it as any disadvantage in the performance of their duties at Colney Hatch, that by the infrequency of their visits, to which allusion is made in your letter, they avoid the danger of becoming reconciled by custom and habit to the continuance of evils that might be removed.

There is no doubt that the Lunacy Commissioners were successful in achieving their own objectives. They were against physical restraint. They set general standards for the quality of life in asylums—food, water, warmth, and clothing. But they failed in two ways. They were criticized for not doing enough to prevent wrongful confinement. This was a middle-class concern. The notion that eccentric individuals could be—and were—locked away as lunatics so that their relatives could get their money was a common one. It was supported by activists and a pressure group, the Alleged Lunatics Friend Society, founded in 1845, and by sympathetic press coverage in *The Times* and the *Daily Telegraph*. From time to time there were lunacy panics. Books were published. In *Hard Cash* written in 1868 by the campaigning novelist Charles Reade, the hero is confined in private mad-houses by his father who wants his money. Samuel Gaskell is portrayed as 'Dr Eskell', a well-meaning but ineffectual bureaucratic bungler. It is doubtful whether the concerns about malevolent incarceration addressed by Reade in his book amounted to very much; they happened—but were rare. There was more substance to his comment in it that 'Inspectors who visit a temple of darkness, lies, cunning, and hypocrisy, four times a year, know mighty little of what goes on there the odd three hundred and sixty-one days, five hours, forty-eight minutes, and fifty-seven seconds.' This was the second, and much more important failure of the Commissioners. Even in otherwise well-run asylums they knew that patients were physically maltreated by attendants. They did their best by looking for black eyes on their visits, but knew that it was something that could not be inspected out. A landmark case was the trial in 1870 of two attendants, Wood and Hodgson, at Gaskell's own asylum, Lancaster, for killing a patient.

William Wilson was about 50 years old and was admitted to the asylum on 15 December 1869 suffering from general paralysis. Because he was restless and sleepless on his first night he was transferred to the refractory ward. What happened the next day was described at the trial by another patient, Dutton.

> I remember William Wilson, and have seen him in No. 1 Ward in the Asylum. I cannot say how long it is since I saw him. I saw Hodgson commence boxing and larking with Wilson, and afterwards he struck him heavily, and knocked him down, and kicked him. He kicked him several times in several parts of his body. When Hodgson had finished with him, I persuaded Wilson to come and sit beside me, and as we were going away I heard the prisoner Wood say, 'Wait till my nose has done bleeding, and you'll see what I will do for you.' I did not see Wilson strike Wood on the nose, as my head was turned the other way. I tried to persuade the patient to be quiet, and told him if he interfered with the attendants he would get ill-used. Wood then proceeded to wrap some towels up, and he afterwards went up to Wilson, got hold of him, and struck him in different parts of his body. Wood then pulled him down on the boards, held him by the collar, and kicked him in the

stomach. The patient was then on his knees. Wood then jerked his knee into Wilson's side, while he held him by the collar. He jerked his right knee into the patient's left side. He was holding the patient very tight by the collar. I wish to speak the truth, and if there are any doubts they (the prisoners) have them. I don't wish to say anything that I don't know.

It was alleged that this happened on the Friday afternoon when Hodgson and Wood were the only attendants on the ward. On Sunday the night attendant noticed that Wilson had difficulty breathing. He died of pleurisy a week later. At post-mortem there was an enormous bruise over the upper abdomen and over the chest. The 2nd, 3rd, 4th, 5th, 6th, and 7th ribs on both sides were broken, the 3rd, 4th, and 5th on the left being broken in two places. The jury convicted Hodgson and Wood of manslaughter and the judge sentenced them to seven years penal servitude.

As John Walton in his essay on the Lancaster Asylum has said 'behind the bland façade of the official reports, the asylum was effect-ively ruled by the cunning of the attendants, supplemented by force when necessary, and sometimes giving way to unpleasant teasing and gratuitous violence'. He quotes the ward supervisor's description at the trial of what made an effective attendant: 'The success of the attend-ants with the patients depends on their humouring their peculiarities.' My grandfather held the same view. After working as a clerk for the Railway Clearing House and an engine cleaner on the Midland Railway at Carnforth in North Lancashire he joined the Lancaster Asylum in 1893 as an attendant, ending his career there as Chief (Figure 5.4). He started work in the refractory ward. Since Wilson's death it had moved to new premises and had been renamed 'B' ward instead of 'Ward 1'. He wrote an account of it for me not long before he died. Describing his first days there:

Nothing seemed natural or normal, really it seemed to me another world, with a stage occupied by a peculiar collection of actors, giving a most unusual performance with the force and madness of the drama. There were real fights, in fact many fierce engagements taking place all at one time. Now it was that I learned my first lesson, a sight which impressed me very much, in fact one could have imagined this scene was enacted solely for the purpose of my benefit. I refer to the actions of the attendants in their dealings with this very refractory crowd of patients, who in a very short space of time created a wonderful transformation scene, from pandemonium to perfect peace—the only weapon they used was tact.

The enormous size of asylums made them difficult to manage. Inspectors were not the only people who did not know what happened on the wards. Their medical superintendents were often in the dark as

Figure 5.4   The author's grandmother, mother, and grandfather pho-tographed in the early 1930s outside his asylum house, 'The Lodge', Lancaster.

well. Even obvious indicators of violence went unnoticed or were explained away because of a reluctance to accept that things were not as they should be. 'Asylum ear' is the classic case. This condition went under a variety of names—haematoma auris, othaematoma, and insane ear. Many papers, books, and theses were written about it in the last half of the nineteenth century. There was agreement that the condi-tion started as a large bruise on the external part of the ear, usually at the front, and that a deformity like the cauliflower ear of boxers, wrestlers, and football players was the end result. It was commoner in restless and excited patients and was also seen in the demented and the mentally retarded. It was commoner in men than in women—a survey at the West Riding Asylum (later Stanley Royd) in the 1870s found it in 3.39 per cent of male patients but only 1.11 per cent of females. Its presence carried grave prognostic significance. Bucknill said in his series of papers 'On the Pathology of Insanity' published in the *Journal of Mental Science* in 1858 that 'it occurs in the worst and most hopeless cases'. Despite its close similarity to pugilists ear, its particu-lar commoness in troublesome patients, its occurrence only on the outside and not on the inside of the ear, its increased frequency on the left—the side that would be naturally struck by a right-handed assailant—and a report of an othaematoma epidemic which ceased after the dismissal of certain attendants who had been ill-using patients, asylum doctors found it difficult to accept that it was just caused by blows. They studied its pathology at post-mortem and claimed that patients were disposed to it because of a 'dyscrasia peculiar to the insane', or 'peripheral trophic' changes or a 'disturbance of the cervical sympathetic nerves'. Some said that alterations in small blood vessels analogous to those in cerebral apoplexy were responsible. Degenerative changes in cartilage to which the insane were particularly disposed were also blamed. Others thought that the increased density and compactness of the skull bones in lunatics obstructed the blood flow and caused a predisposing oedema. A suffusion of blood to the ears during excitement coupled with slight trauma was held to be responsible. Investigators who isolated bacteria from effused blood of recent cases said that they played a significant role. An appropriately named doctor from Nottingham, W. Phillimore Stiff (he changed his name to Phillimore W. Phillimore in 1874) and a Russian, W. P. Tischkov, claimed that it occurred spontaneously. But the evidence is overwhelmingly in favour of the view widely promulgated in the early 1860s by the German psychiatrist Bernhard von Gudden. He said that it was entirely due to maltreatment. He showed in his papers that deformed ears identical to othaematomas of the insane also occurred in the sane. They had been described by classical Greek writers and could be seen on ancient statues of Mars, Hercules, and Pollux. When he wrote these papers he was in charge of an asylum at Werneck in Bavaria. He had built a substantial reputation as an organizer by creating it in the baroque buildings of a former Episcopal palace. In his asylum he showed that by holding attendants responsible, othaematoma almost disappeared. In his practice Gudden gave his patients as much personal freedom as possible. Constant contact between doctors and patients and well-trained staff were the norm. He went on to become one of the leading German psychiatrists. His experimental work on the anatomy and function of the brain, particularly the parts concerned with sight, gave him an international reputation. His name still appears in modern textbooks of neuroanatomy. But his views on 'asylum ear' did not seem to be particularly influential in Britain, despite being abstracted in English in places like the *British Medical Journal* and the asylum doctors own publication, the *Journal of Mental Science*. Even the dramatic circumstances surrounding his death made no difference. King Ludwig II of Bavaria had been mad for years. He was a great friend of Richard Wagner and used to dress up as Tannhäuser, Tristan, and Lohengrin. He had a reservoir and wave machine constructed on the roof of his palace at Munich on which he would sail in a boat with a stuffed swan floating in front.

Figure 5.5   Montage of King Ludwig II and Neuschwanstein.

To suit his whim the water was coloured blue with copper sulphate. It leaked into the palace and ruined its lavish furnishings. He had hallucinations. He would bow to particular trees and bushes and order special trains so he could visit pieces of architecture that he had read about. He spent vast sums building castles, including the fantastic Schloss Neuschwanstein (Figure 5.5). In March 1886 Gudden was consulted. With three other psychiatrists he concluded that Ludwig was suffering from incurable paranoia. A Regency was declared on 9 June and on the next day Gudden and others went to Neuschwanstein to tell Ludwig. The King's coachman had given him advanced warning and the arrivals were arrested. He ordered that their eyes be put out and the flesh be torn from their bones. A little later Ludwig's gendarmes learned of the Regency and released them. On 11 of June Ludwig was taken to Schloss Berg, on the Starnberg lake, 20 miles south of Munich. On Sunday, 13 June he took a lunchtime walk with Gudden, appearing quiet and friendly. They took another walk at half-past six in the evening. Two keepers were sent after them, but the King sent them back. At eight o'clock the King had not returned and a search began. Ludwig's stick and Gudden's hat were found near the lake, and at 10.30 PM their bodies were found floating face down in the shallows, 20 paces from the shore. Gudden had scratches on his nose and a bruise on his forehead. The King's watch had stopped at 6.54 PM.

The official view was that Gudden had been murdered by Ludwig, who then committed suicide. After all, he was paranoid as well as being 6 ft 3 in. tall and 20 years younger than the psychiatrist who had played a major role in the removal of his regal powers. Conspiracy theorists have come to different conclusions that need not concern us here. Suffice it to say that although Gudden was criticized at the time for not being more careful, his British obituarist in the *Journal of Mental Science* pointed out that 'his name must be added to the not inconsiderable list of physicians who have been either injured . . . or killed outright by patients of whom they have had charge'. Although Gudden's violent death was described as being 'known throughout the civilised world', it had the opposite effect regarding his conclusions about asylum ear. In a definitive review on the subject published in the *Edinburgh Medical Journal* in 1894 his name was not even mentioned.

It is a pity that in their inspections the Lunacy Commissioners did not use the approach adopted by the World Health Organization (WHO) when it surveyed countries to find out about poliomyelitis. For most nations data from health ministries were useless. Nobody responsible for controlling disease wanted to admit that they had a problem. Officials did not collect data on the incidence of disease. Because there was no record of the number of cases they could truthfully say that there was no evidence of a problem. It was far easier and far more revealing for WHO to assess polio in a country by counting the number of children in the streets with withered limbs; counting cauliflower ears would have been an equally good measure of the level of violence in a ward and in an institution.

Whatever the deficiencies of the Lunacy Commissioners, there is no doubt that their activities brought real benefits to asylum inmates. An important reason for this success was that some of them, like Gaskell, were single-minded zealots. They had a passion to do good, but were not hypocrites. Their energy was inexhaustible and they were incorruptible. The prestige of their position significantly helped their authority. Gaskell's salary as a Commissioner put him into the top 10 per cent of doctors ranked by earnings. He earned more than three times as much as the medical superintendents of the asylums he was inspecting. Being zealots counteracted the possibility that their intimate knowledge of the asylum scene and friendship with those they were inspecting would lead to regulatory capture, which, if they had been indolent, could have led to the sort of inspections described so graphically for prisons—with their internal inspectorate—by Frank Norman in *Bang to Rights*.

> Every now and again this nick like all other nicks in the country, would get a visit from the commissioner. Now this is a very big day in the life

of screws and the governors, and the day before he is due to arrive they start haveing a big clean up, all the floors are scrubbed and the brasses are polished, also the recesses are give a good clean out. The exersize yard is swept and the flower beds are weeded and we are told to make sure our peters are clean and tiedy.

At last the great day is here. The Commistioner usualy wears a black over coat and black Antoney Edden hat even in the summer, he also quite often carrys an umbreler. I don't realy know why they make such a fuss of him as all he does is come and have lunch with the governor have a little look around the nick get in his big car and drive off again. Some times he comes round just as we are being fed, and may pick on some one and say. 'What's the food like? Any complaints?'

You look at the governor and the chief in turn (for they are his constant companions while he is in the nick) you then answer. 'No complaints; sir.'

Zealous commissioners could also see through the sort of devices that my grandfather used to soften scrutiny:

Whenever it has been my duty to escort a member of the committee through the wards, and should I discover that he was full of grumbles, and on the outlook for faults, I usually conducted him to the refractory ward, and found my august visitor soon requested me to show him the nearest way out.

The Lunacy Commission was replaced by the Board of Control in 1913. This was in turn abolished by the 1959 Mental Health Act. Inspections continued but focused on institutions that contained patients detained against their will. The problems encountered by the Lunacy Commissioners in finding out what actually happens on wards were not—and have not—been resolved by these changes. The comments in the 1992 Report of the Committee of Inquiry into Complaints about Ashworth Hospital (a forensic psychiatric hospital at Maghull near Liverpool) relating to the work of the Mental Health Act Commission (MHAC) could have been written by Samuel Gaskell:

Everything we have heard and read about in the course of our Inquiry leads us to the broad conclusion, that at Ashworth—and we speak only of that Special Hospital—the MHAC has not been nearly as effective as its regularly visiting would be expected to produce. Much of the sub-standard patient-care which we describe . . . appears to have gone unobserved by the MHAC.

We say unobserved in the sense that the regimes of ward-life went unnoticed. In part the reason may have been that Commissioners tended to spend the time on their visits, justifiably, closeted with patients and talking to staff about patient-problems, as well as checking hospital documentation. Moreover, questionable activities on the ward would inevitably tend to occur when no Commissioner was in sight.

Commissioners were never able to visit unannounced; even had that been possible, an unannounced visit would not have revealed the untoward incident. It may also be that the occasional outsider coming inside the hospital, even if irregularly performing a strictly visitorial function, sees only the superficial and relies heavily on the reporters of incidents. Nothing short of an audit or a permanent presence on the hospital site can reveal the reality of life on the ward.

The Lunacy Commissioners and their modern successors clearly had problems in finding out what was going on at ward level. But when inspections did reveal things that required action, were their powers sufficient to make things happen? On 23 July 1870 *The Lancet* published a leader on the Lunacy Commissioners. It commented in particular on rib fractures in asylum patients. It said that in their recently published 24th Report, the Commissioners did 'not even refer to the fragility hypothesis'. The journal clearly implied that the Commissioners were sceptical about the hypothesis, which claimed that many lunatics were particularly susceptible to broken ribs because they had soft bones, or brittle bones, or osteomalacia. On top of bone fragility it had also been suggested that because of the slothfulness of nerve currents the 'excito-motory acts tardily and therefore muscular contractions are too late to guard against the injurious effects of falls and blows'. The fragility hypothesis is the asylum ear story with its special pleading and 'insane diathesis' excuses all over again. The death of William Wilson later in the year at Lancaster was to prove the Commissioners to be right. They knew that patients were being assaulted and that this explained why lunatics got broken ribs. So instead of accepting the validity of exculpatory pathological speculations the Commissioners were agreeing with the views of Gudden. In his paper 'Ueber die Rippenbrüche bei Geisteskranken' in the 1870 volume of *Archiv für Psychiatrie* he had followed up his studies on asylum ear by looking at broken ribs and had come to the same conclusion, that maltreatment by attendants was sufficient to explain everything. Continuing its discussion of the Commissioners Report, *The Lancet* went on

> they place on record that during 1869, out of 89 male attendants dismissed . . . 35 were dismissed for assaults on patients . . . and they dwell with perfect propriety and much force on the innumerable small miseries short of broken bones . . . that a brutal attendant may inflict day after day and hour after hour upon the helpless creatures under his charge. The law which requires the Commissioners to investigate and report as to matters affecting the management of country asylums has invested them with no authority to enforce their own.

In essence the Commissioners were being criticized for being aware of a problem but responding by the wringing of hands because Parliament had not given them enough power to intervene. There is no doubt that the issue of who to blame when asylum lunatics suffered fractures continued to be vigorously debated for many years after the Commissioners' 1870 report. A particularly interesting development was the invention of an instrument to automatically measure the 'breaking strain of the ribs of the insane' by the London psychiatrist, Charles Mercier. He had a number of them made and sent to asylums in the hope that they would be used to establish that the sort of patients that got broken ribs already had softer and weaker bones—maybe even to the point that fractures occurred spontaneously. The instrument was used *post-mortem*, of course. The response to Mercier's initiative was not good, however, although data was collected which undoubtedly showed that some patients in asylums had bone problems. Nowadays many would probably be diagnosed as suffering from osteoporosis. Mercier's motives were probably mixed. He had a particular interest in medico-legal matters and no doubt hoped to protect asylum staff from unjust accusations. But his interest in soft bones was a personal one. He had them himself. At about the time that he invented his instrument in the mid-1890s he began to develop Paget's disease, a disorder in which bone turnover increases leading to a disorganization and weakening of its structure. His head became very large because of skull thickening and he became almost completely deaf because new bone formation compressed his auditory nerves. His disease killed him in 1919 (Figure 5.6).

The issue of the powers of inspectorates was an important one at Stanley Royd. The kitchens there were regularly inspected by environmental health officers. Every year through the early 1980s their reports contained adverse comments. The poor design of the kitchen as a whole and the high temperatures in the 'long store' were mentioned frequently. But most of the negative comments related to walls, floors, and ceilings. The final report before the outbreak said that the hospital should

> 1. Replaster the area of perished wall plaster in the bread store. 2. Provide a smooth impervious hard wearing surface to the floor of the walk-in refrigerator. 3. Paint the bare wooden door to the pan wash to provide an impervious readily cleansable surface. 4. Provide a suitable storage facility for cleaning materials in the butchery section. 5. Replace the rusty racks in the butchery refrigerator and provide it with a smooth impervious hard wearing floor. 6. Replace the cracked water closet seat to the gentlemans water closet in the store corridor.

Florence Nightingale would have been proud of this report. She would have said that implementing its recommendations would have made floors, ceilings, and doors less pervious to 'excremental exhalations'. But we now know that it would do absolutely nothing to keep

Figure 5.6   Charles Mercier in the late stage of Paget's disease.

Salmonella at bay. As with Barr's, a more detailed and thorough inspection was done immediately after the outbreak and similarly it revealed a long list of major deficiencies not detected—or reported on—shortly before it. These included the absence of effective and adequate temperature controls, the inadequate cleaning of working surfaces, a failure to protect food from contamination, and a lack of food hygiene awareness by the staff. There was 'a general appearance of restrained chaos mitigated by the number of long term employees who are familiar with the organisation'. The pre-outbreak report had missed all the deficiencies that played key roles in causing it. It was a faulty inspection because the things that it missed had not been hidden, or were sporadic, but had been obvious. One reason for its superficiality was that it had been done by an officer on the last day of his employment. It had been a very busy day for him. In four-and-a-half hours he had visited five hospitals because he did not want to leave unfulfilled commitments for his successor. But a much more important reason lay in the attitude of environmental health officers to hospitals and their perception of the enforcement powers they had over them. In 1984 hospitals enjoyed Crown Immunity. Because their premises were owned by the Crown and because their employees were Crown servants, they could not be prosecuted. On account of this environmental health officers felt that their role was essentially that of visitors and advisers. They paid less attention to detail than they would in commercial premises and left things to hospital doctors and catering managers. The Stanley Royd public inquiry considered this situation at length. Their conclusion about Crown Immunity as an excuse for a hands-off approach was the same as that opinion of Sheriff Cox and Lord Cullen about the 'inspection is only a snapshot on the day' defence put forward by North Lanarkshire Council and the Department of Energy. The public inquiry was not impressed. In particular they pointed out the availability of the 'Crown Notice'—a stern document which said that although failure to comply with an instruction could not attract a prosecution it would 'result in a formal approach...to an appropriate person with higher authority...if necessary...to the responsible Minister'. At the end of the day the important thing was compliance, not prosecution.

National Health Service (NHS) hospitals no longer enjoy Crown Immunity. But some environmental health officers still complain about the inadequacy of their legal powers. Examination of the legal structure that underpins their inspectorate gives substance to this argument. It is regarded by the judiciary as dealing with 'quasi' crime—'mala prohibita'—rather than 'real' crime. Its laws are statutory, unlike much of criminal law, which is uncodified and derived from cases. The law itself is often exhortatory and the 'due diligence' defence mitigates strict liability. Cautions are usually given before enforcement. In practice, legal action is often only taken after an outbreak, and penalties seem to fail to reflect the harm that has been caused. The outcome of the criminal trial that followed the Central Scotland *E. coli* O157 outbreak is typical. For breaches of hygiene under the Food Safety Act and Regulations, and for selling meat contaminated with *E. coli* O157 the John Barr partnership was only fined £750 on the first charge and £1500 on the second. But even if environmental health officers are right to compare the attitude of the judiciary—and some in the food industry—to the laws they enforce with the way the public regards speeding or driving with underinflated tyres, history is not entirely on the side of those who complain about the insufficiency of legal powers. One early inspectorate had no legal right to prosecute but achieved many successes in the face of determined and powerful opposition. Pressure to regulate the rapidly expanding railways in the late 1830s led to the establishment of parliamentary committees which recommended the setting up of a regulatory authority. In consequence the Railway Department of the Board of Trade started work in 1840. Its inspectors were authorized to inspect, and were charged with approving, new passenger-carrying railways. Railways were obliged to report to the Board accidents 'attended with serious personal injury to the public'. With only one early and unimportant exception, for almost 150 years all the inspectors were officers of the Royal Engineers. Using soldiers rather than ex-railway employees meant that while the inspectors had technical expertise they were protected from regulatory capture. They were best known for their accident investigations and their crucial role in driving safety improvements from them. But their legal powers were very limited. They only received statutory authorization to look into accidents in 1871. But they interpreted their remit vigorously and broadly and started accident investigations even before the 1840 Act establishing them had come into force. They reported their findings to the Board of Trade. In 1854 these reports were given to the press and in 1860 they were put on public sale. The military Railway Inspectors have had a good press. Their 'naming and shaming' policy pushed railway companies into a series of safety improvements such as the fitting of passenger trains with powerful continuous brakes that came on automatically when a train broke in two. They fought battles with hard-hearted autocratic powerful capitalists, like Richard Moon, Chairman of the London North Western Railway, who said that 'these mechanical appliances were all inducements to inattention on the part of signalmen and drivers', and they held their ground. Their success have been recorded with praise in books like L. T. C. Rolt's minor classic *Red for Danger*.

Two important factors aided the work of the Railway Inspectors. First, many causes of accidents, like train collisions, axle, crank-shaft and rail failures, and boiler explosions, had technical remedies. Mechanical failures could be addressed by metallurgical science, and human error and incompetence could be minimized by engineering approaches like improvements to signalling. Secondly, many of the technical improvements were, for all practical purposes, irreversible. They led to permanent safety gains. The gas-lit passenger coaches responsible for horrendous fires after crashes have gone for ever. But it took events like Quintinshill to drive these improvements. Located on the West Coast main line, just north of the Scottish border near Gretna, its signal box controlled two 'lay by' loops. At 6.30 AM on 22 May 1916 each was occupied by a goods train. As in the lead up to the R101 and Piper Alpha disasters the shift had just changed and the day signalman was busy filling up the train register book with retrospective entries to cover up his late arrival and chatting about the war news in the newspaper he had brought with the night signalman. The Scotch express from London was running late and to allow its passengers to make connections a local train from Carlisle had been given preference. The original intention was to shunt it into a loop to let the express pass. Because the loops were occupied it was switched across to the other main line, where it stopped. The express to the north was given signals for a clear road at 6.38 AM. So far, so good. But errors in the signal box then took over, when four minutes later, similar signals were given for a fast running troop train special, heading south, which had passed through Lockerbie, 11 miles away, at 6.32 AM. At 6.48 AM it burst on the scene running at 70 mph on a downhill gradient, and crashed head on into the stationary local train. Its fifteen coaches with a combined length of 639 ft were reduced instantaneously to a wreck 210 ft long. The engine came to rest across both main lines. Almost at once the Scotch express came into view. It weighed over 600 tons and was being pulled by two engines at high speed. It braked but smashed into the wreckage at 60 mph. The gas cylinders of the troop train carriages had been fully charged just before it started its journey north of Glasgow. Burning coals from its smashed engine set their pressurized contents alight and the mass of wood, bodies, and coal from the engine tenders became an enormous funeral pyre which burned for more than 24 h. At least 227 people perished. The precise number was never established because of the fire and because the regimental roll of the soldiers on the special, the 7th Royal Scots, was destroyed in it. Eighty-two bodies were recovered but could not be identified and 50 soldiers were listed as 'missing' by the Army (Figure 5.7).

So it could be said that the 'learning from experience' and 'blaming and shaming' techniques for improvement used by the Railway Inspectorate required particularly bad experiences—ones nearly always dependent on the loss of many lives—to be successful. Another weakness of this approach was its slowness. One reason for this was the infrequency of accidents horrible enough to drive public and parliamentary opinion to action. Inadequate brakes provide classic examples. A high-speed derailment which killed eight in 1880 led Colonel Yolland, the investigating inspector, to say:

> It is all very well for the Midland Railway Company now to plead that they are busily employed in fitting up their passenger trains with continuous brakes, but the necessity for providing their passenger trains with a larger proportion of brake power was pointed out by the Board of Trade twenty years since; and with the exception of a very few railway companies that recognized the necessity and acted upon it, it may be truly stated that the principal railway companies throughout the kingdom have resisted the efforts of the Board of Trade to cause them to do what was right, which the latter had no legal power to enforce,

Figure 5.7   Quintinshill: the troop train burns.

and even now it will be seen by the latest returns laid before Parliament that some of those companies are still doing nothing to supply this now generally acknowledged necessity.

The final crunch came in 1889. The Newry to Armagh branch of the Great Northern Railway of Ireland was old-fashioned. Trains were dispatched up the steep three-mile gradient from Armagh on a time interval basis with not less than 10 min between passenger trains and 20 min for a passenger train after a goods train. They were fitted with Smith's non-automatic vacuum brakes; if the brake pipe broke the vacuum was lost and the brakes stopped working. On 12 June 1889 a special Sunday school excursion to the seaside at Warrenpoint was organized. When the driver, Thomas McGrath, arrived from Dundalk he found fifteen carriages with 941 passengers. Following the usual practice with children's excursions, to keep everyone in order the carriage doors were locked. McGrath was unhappy. He had never driven on this road before. He did not think his engine was powerful enough. He sought out John Foster, the station master, and told him that Dundalk had said that his maximum load was thirteen vehicles. Foster accused him of grumbling and they had words. In bad humour McGrath started the train up the hill. It stalled near the top. James Elliot was in charge of the excursion and was travelling on the footplate. He and McGrath decided that the best thing to do was to divide the train and take the first five carriages over the hill, and then come back for the rest. But when this happened the vacuum brakes of the carriages were lost. In anticipation, the wheels of the rear coaches had been scotched with stones and the guards van handbrake had been screwed down.

When McGrath eased back a little after uncoupling the 5th and 6th carriages, all this was for nought. The wheels rode over the stones and the rear ten carriages, containing 600 people, started to roll back down the hill. Meanwhile another train had left Armagh and was steaming up the hill at 30 mph. The driver saw the runaways approaching and managed to slow down, but there was a tremendous collision. Three carriages from the excursion were completely destroyed. Eighty people, many of them young children, died, and 260 were seriously injured. The shock to public opinion was tremendous. It propelled regulatory legislation at unparalleled speed. Just over six weeks later, on 30 August, the Railway Regulation Act was passed. On 24 October the Board of Trade acted under the powers given it by the Act and told railway companies that the absolute block system, which meant that a train could not enter a section until the previous one had left, the interlocking of signalling and points, and automatic continuous brakes were now compulsory. Companies were given one year to introduce block working, and 18 months for the adoption of interlocking and continuous brakes.

It is true to say that the success of the Railway Inspectors was significantly tempered by the slowness with which railway companies adopted their recommendations. As with other policy makers they needed disasters to help them. But the inspectors were seen to be successful in reducing the amounts of dramatic accidents in spite of eschewing legal enforcement powers. Indeed, the Railway Regulation Act was the only safety legislation passed in the nineteenth century. However, examination of their record in preventing *all* the deaths and injuries on the railways shows a much more indifferent record. Take shunters. They had a high mortality rate from being crushed between wagons that they were coupling or uncoupling. The Royal Commission on Accidents to Railway Servants in 1900 found that their accident rate was worse than for miners. The only technical improvement that had been introduced in the whole of the nineteenth century was the shunting pole, a metal tipped stick supplied to railway workers from the late 1870s to allow coupling and uncoupling to be done from the side of the wagon. Unions pressurized companies to introduce couplings that did not require guards and shunters to endanger themselves. The Competitive Coupling Trial held by the Amalgamated Society of Railway Servants at the Nine Elms goods yard of the London and South Western Railway in 1886 is a good example of their work. C. E. Stretton in his book *Safe Railway Working* summarized its report. It opened with the statement that in 1884, 130 men were killed and 1305 injured in shunting operations. It then announced the award of six prizes to inventors of automatic and non-automatic couplings that could be operated from the side of the wagon, two of £100, two of £50, and two of £25. But in practice nothing changed. The main reason was that the cost of introducing automatic couplings would not only have been vast, but that it would have fallen in large part to private companies. The scale of the problem and its persistence was shown more than 60 years later at railway nationalization in 1948 when the newly formed Railway Executive inherited about 1.25 million wagons of 480 different kinds, most of a design going back to the 1830s with hook-and-chain couplings, individual hand brakes, and grease axle-boxes. Many were very old. They had survived because they suited the many private sidings with their small size and sharp curves. This type of wagon persisted until the 1980s. Automatic devices like the central buckeye knuckle-like coupling only began to be fitted in numbers in the 1950s. The contrast with other countries is stark. The Janney buckeye was adopted as standard in the US in 1888 and made compulsory by Congress—with a seven year introduction time—in 1893. Despite an increase in railway employees from 873,602 in 1893 to 1,189,315 in 1902 the number of shunting accidents fell from nearly 11,000 to just over 2000 during that decade.

Not all accidents to railway workers were preventible by technical solutions. My grandfather's experience when he worked as a wagon number taker for the Railway Cleaning House was typical.

In April 1890 the accident happened when I was on duty from Sunday to Monday morning. I booked on at 9 AM on Sunday and on the same sheet booked off at 6 PM. On another sheet I booked on at 6 PM Sunday and signed off at 7 AM on Monday. It was all continuous duty.

In the early hours of Monday morning a London North Western goods train ran into the goods road. To see if there were any wagons to detail for the Furness Railway I crossed the lines. The engine began to shunt and I retreated towards the office. As I was crossing the line over which the engine and wagons must pass, the heel of my boot caught fast between two converging lines. Immediately I saw my great danger for I could hear the engine coming. I tried to unfasten the lace of my boot but I was wearing strongly laced boots and it was in vain. I saw that the engine would be on me if I could not escape in more than a minute, but my shouts were not heard against the noise of the engine. When the first wagon was only a few yards away and I had almost given up hope, to my great joy the heel of my boot came off and I stumbled forward out of harms way. After a while I made my way to the office when I had a good rest then continued my duties until relieved at 7 AM.

Whether the very long hours that my grandfather was working had anything to do with his brush with death is impossible to say. But they were universal at that time. It is clear from contemporary Railway Inspectorate records that they played an important role in the causation of accidents. The inspectors commented on them adversely on many occasions. Change came only slowly. This is not surprising, because similar patterns of work occurred in other industries. My grandfather found this out when he started work at the Lancaster Asylum in October 1893 as an attendant at £28 per year with board, lodging, laundry, and uniform. He worked on average an $86\frac{1}{2}$ hour week. Every month he had two days free and two short shift days, one of 8 h and one of 10 h, counterbalanced by two 15-h days. Of the remaining days, on nine he worked 13 h and on twelve he worked 14 h. He slept on the ward near the padded cell but got very little sleep because of the raucous and disturbed patients. In his own words there were times when he 'longed to be in the goods yard, among the rattling wagons, or in the engine shed where the oil and grease ran through your clothing, and down your arms and legs.'

Ever since their invention railways have been seen by the public as an important source of danger. This is paradoxical. Even before the establishment of the inspectorate they were much safer per mile travelled than any other form of transport. On the basis of the number of deaths in accidents and miles travelled, Dionysius Lardner calculated that in 1848 that the chances of a passenger suffering an accident fatal to life on the railways of the United Kingdom was 65,363,735 to 1. Railways are much safer now than in Lardner's time. But the negative perception about railway safety is as strong today as it ever was. It feeds through into government expenditure. Currently, it spends about £0.1 million to prevent each road fatality. The cost per fatality avoided by the Train Protection Warning System (TPWS), which is in the process of installation and will stop a train within the overlap between a red signal and the junction it is protecting—so long as the train is moving at less than about 70 mph—is more than £10 million.

Lardner was the first to explain this paradox. In an article on railway accidents published in 1859 in his multivolume work *The Museum of Science and Art*, he wrote:

> In the modes of travelling used before the prevalence of railways, accidents to life and limb were frequent, but in general they were individually so unimportant as not to attract notice, or to find a place in the public journals. In the case of railways, however, where large numbers are carried in the same train, and simultaneously exposed to danger, accidents, though more rare, are sometimes attended with appalling

results. Much notice is therefore drawn to them. They are commented on in the journals, and public alarm is excited.

This is the 'Lardner effect'. Its explanatory power is just as important in food safety as in transport. By any criteria—the degree of public awareness, column inches in newspapers, the number of TV specials, the frequency of government inquiries and task forces, and the number of special interest groups—*E. coli* O157 wins hands down against *Campylobacter*. And yet reported *Campylobacter* infections are about 50 times commoner than *E. coli* O157. *Campylobacter* causes diarrhoea in people far more often than all other food poisoning bacteria put together. But *E. coli* O157 is a common cause of newsworthy outbreaks, and *Campylobacter* is not. One in ten cases of *E. coli* O157 are outbreak victims. For *Campylobacter* the proportion is one in five hundred. *E. coli* O157 outbreaks are also noteworthy because they often affect young children. Serious complications are not uncommon and some patients die. *Campylobacter* is much less dramatic. It hardly ever kills. Equating outbreaks to rail crashes provides the Lardnerian explanation as to why the public is hardly aware of *Campylobacter* and why most doctors care even less despite its commonness.

Lardner's understanding of how the public assesses risk is a particularly modern one. This is not surprising, because despite being born in 1793, his career has many modern features. Nowadays academic scientists are not only supposed to engage with industry, interact with society at large and explain their activities to the general public, but work hard as well, particularly by publishing learned articles and books that will stand the test of time. Lardner did all these things (Figure 5.8). In 1828 he was appointed the first professor of Natural Philosophy (Physics) at the newly founded University College, London. He did experimental work on the Great Western Railway when Brunel was in charge. He hobnobbed with aristocrats and important people like Bulwer, the writer, and Disraeli, being satirized by Thackeray (as Dr Ignatius Loyola, and Dr Dionysius Diddler) for his pains. In 1831 he resigned his chair because the University was in financial difficulties and because he could make more money by writing and lecturing. His main works were the *Cabinet Cyclopaedia* of 133 volumes, and *Railway Economy*, published in 1850. Much cited, it has lasted well. It was reprinted as an 'economic classic' in New York in 1968. His reputation as a popularizer of science was very great and it was said that in the 1840s his lecture tour of the USA and Cuba made £40,000. He had close relationships with the media and was rumoured to be the Paris correspondent of the *Daily News*. The private as well as the public lives of modern academics have been satirized most acutely in David

Figure 5.8   The Revd Dr Dionysius Lardner.

Lodge's novels; Lardner would fit in them very well. As the *Dictionary of National Biography* says,

> in the midst of these various and arduous labours Lardner carried on during several years an amour with Mrs Heaviside, the wife of Captain Richard Heaviside, a cavalry officer, and eloped with her in March 1840. Heaviside obtained a verdict against him in an action of seduction, with £8,000 damages. An Act of Parliament dissolving the marriage followed in 1845.

Public concern about rail safety is as acute now as it was in Lardner's time. It is still fuelled by accidents. Proof of how seriously it is taken by government is contained in Lord Cullen's preamble to the terms of reference of his inquiry into the head-on crash at high speed on 5 October 1999 at Ladbroke Grove, two miles west of Paddington, which killed 31 and injured more than 400. 'As a consequence of that crash I was appointed on 8 October 1999 by the Health and Safety Commission, with the consent of the Deputy Prime Minister, to conduct a Public Inquiry....' His report makes it clear that the Railway Inspectorate could not be absolved from blame. Its parent organization, the Health and Safety Executive (HSE), admitted that there was a lack of resources, that there was a lack of vigour in pursuing issues, and that it placed too much trust in duty holders. Not only was it assuming compliance from Railtrack that wasn't there, it didn't know as much about what was going on as it should. In the words of Jenny Bacon, Director-General of the HSE: 'on one or two occasions it is wool pulled over the eyes, more generally it is not being in touch with what has been going on . . .' Her words would have done very well to explain the relationship between inspectors and Mr Barr before the Central Scotland outbreak.

It is clear that the difficulties faced by inspectorates are deep seated, general, and persistent. Is this a particularly British problem? After all, the ones that we have been considering evolved in the context of the unique British constitution, and in response to British disasters. The clear answer is no. It was dramatically illustrated by the Walkerton *E. coli* O157 outbreak in Canada in 2000. The parallels of this disaster with Wishaw are uncanny. Once again the organism exploited the ignorance of providers of products to the public—in this case water— and once again it tested regulatory systems and found them wanting. The outbreak first showed itself on Thursday, 18 May 2000, with the absence of twenty children from the Mother Teresa school and the admission of two children to hospital with bloody diarrhoea. Local residents made enquiries about the safety of the Walkerton water supply. A staff member of the Public Utilities Commission (PUC) assured them that it was safe. Repeat enquiries the next day from the local health unit were given the same message. The water was 'OK'. Similar reassurances were given to callers on Saturday. But on Sunday, 21 May, confirmation came that *E. coli* O157 was causing infections, and the health unit issued a boil water advisory notice. At the end of that day the Walkerton hospital had received more than 270 calls about symptoms of diarrhoea and serious abdominal pain. By Wednesday several patients had been flown by helicopter to London, Ontario, for specialist treatment, and four had died. By the end of the outbreak nearly half of the 4800 Walkerton residents had been ill, 27 had developed HUS and a total of 7 had died.

The immediate cause of the outbreak was not difficult to establish. Walkerton's water came from three wells. Well 5 was the main source just before the outbreak (Figure 5.9). It was shallow and drew its water from highly fractured bedrock with a thin layer of topsoil, an ideal design for letting surface water into it easily and rapidly. In late April manure had been spread near it, and between 8 May and 12 May there was heavy rain.

Well 5 water at this time was being chlorinated, but at lower levels than was needed to ensure bacteriologically safe water. Between 13 May and 16 May the organic contamination on the surface, including *E. coli* O157, entered the well as the rain water drained away. It overwhelmed the disinfecting capacity of the chlorine, allowing live organisms to pass into the water supply of the town. That all this

Figure 5.9   Well 5, Walkerton.

happened was no accident, or freak of nature. It was due to a combination of deceit, ignorance, incompetence and regulatory failings
occurring at levels that beggar belief. The Walkerton water system is
owned by the town and operated by the Walkerton PUC. Its general
manager was Stan Koebel, and his brother Frank was its foreman. For
years PUC staff had believed that the water they produced from their
wells was naturally safe. Even before the Koebel brothers took over in
1988, the notion that it needed treatment was treated with disdain.
There was a host of improper operating practices, which Stan Koebel
continued. Wells were operated without chlorination, false entries
were made in daily operating records, chlorination was being done
inadequately, the daily measurement of residual chlorine was not being
carried out when it should, false locations were given for samples sent
for microbiological testing, and false reports were being sent to the
regulatory authority, the Ministry of the Environment (MOE). So
when Frank Koebel checked Well 5 on 13 May, he did not measure the
chlorine residual—a test of whether its disinfecting power was being
overwhelmed by contamination. It wasn't done on the 14th or the
15th either. Routine microbiological samples were taken on the 15th by
another PUC employee. They were deliberately mislabelled. Although
it is almost certain they came from the PUC workshop, downstream
of Well 5, they were labelled as coming from 'Well 7' and 'Durham
Street'. Two days later, phone calls and faxes to Stan Koebel from the

private laboratory doing the tests said that they showed gross contamination. But for the next two days he continued to reassure callers, including the health unit, that the water was safe. There were at least two reasons why he did this. Since 15 May he had been operating Well 7 without a chlorinator. He knew that this was wrong and wanted to conceal it from the health unit. Neither did he understand how nasty *E. coli* O157 infection could be. Although both he and his brother held class 3 water operator's licences, they had both obtained this certification through 'grandparenting', the award of the licence on the basis of long experience, rather than the possession of qualifications obtained through examinations or a period of formal training. The MOE required 40 h a year training for certified operators, a kind of retrospective safety net for the ones that had come through the grandparenting route. Neither brother fulfilled this requirement.

The Koebels worked for the PUC. But its commissioners had very little knowledge of technical matters. They concerned themselves with finance and left everything else to Stan Koebel. They trusted him. In May 1998 they received a report from the MOE which said that *E. coli* organisms (not *E. coli* O157) had been present in a number of treated water samples. There should have been none. Their presence indicates unsafe water. The report emphasized the importance of chlorine residuals, and said that the PUC was not complying with bacteriological sampling requirements and was not keeping proper training records. The commissioners did not ask Stan why *E. coli* was there or why these other things were not being done. They left it to him to sort things out. No follow up checks were done.

The activities of the Koebels were inexcusable. But it would be very wrong to hold them or their employers, the commissioners, solely responsible for the outbreak. The regulatory body—the MOE—was equally culpable. When Well 5 was constructed in 1979 the MOE gave it a Certificate of Approval knowing that it was vulnerable to surface contamination. Over the years it tightened up practice, and in 1994 the Ontario Drinking Water Objectives were amended to say that such vulnerable water supplies should have their free chlorine and turbidity continuously monitored. But the MOE never went back to review the Well 5 approval, or ask Walkerton to install continuous monitors. Even worse, when it inspected the Walkerton water system in 1991, 1995, and 1998, it did not assess Well 5 at all, although information about its vulnerability was in its files. In addition the inspectors failed to spot the improper chlorination and monitoring practices, even although examination of the records would have disclosed them. In 1998 *E. coli* was being detected in the water, chlorine residuals were lower than they should be, and microbiological testing was not being done as often as it should be. The inspectors noted these things. But no follow up inspections were done, and none had even been scheduled when the outbreak struck two years and three months later.

If the bad water test results of the 15 May had been sent not just to the Walkerton PUC but to the MOE and the local Medical Officer of Health as well, the boil water notice would have been issued two days earlier than it was, and there would have been 300-400 fewer cases. This failing was a direct result of government policy. To save money, in 1996 it had stopped conducting water tests for municipalities. This obliged all the small ones to use private laboratories. A guidance document had been issued by the MOE which recommended that a clause should be put in contracts specifying that the laboratory notify the MOE and the local Medical Officer of Health about bad results. But Walkerton neither asked for or got this document. In 1997 the Minister of Health wrote to the Minister of the Environment to sort out the general problem of how to make sure that results would automatically be sent to all those who needed them in order to protect the health of the public. The Environment Minister replied that he would not legislate and that it should be carried forward through the Ontario Drinking Water Objectives—guidelines rather than regulations. Nothing more happened until after the outbreak.

So just as in Wishaw a high-risk operation was being run by untrained and unqualified people who only had their own experience to guide them, people who were ignorant of the risks they were running and ignorant of the measures they should have been taking to reduce and eliminate them. Technical advice was not taken seriously and corners were cut. Years of operation without adverse effects on their customers had given them the confidence to go on working in time-honoured, but as it turned out, catastrophically dangerous ways. The whole point of an inspectorate is to identify this kind of behaviour, and stop it. They failed.

But catastrophes like Wishaw and Walkerton are rare, and on balance inspectorates have been a good thing. Without them the lot of the lunatic would have been much worse and there would have been many more mine disasters and rail crashes. But despite more than a century and a half of learning from experience their routine performance still fell short. They became better at reducing the frequency of major disasters rather than run-of-the-mill problems. Take coal mines. After the Second World War part of Stanley Royd Hospital became a district general hospital—Pinderfields. A spinal injuries centre was established there in 1954. Of the 527 admissions between 1960 and 1969, 62 were coalminers. Injuries due to roof falls—a technical issue close to the hearts of mine inspectors—were still one of the commonest causes of paraplegia.

My own experience in the 1960s as a nurse in a large hospital that had been established 90 years before for the 'idiots and imbeciles of the seven northern counties' illustrates the basic challenge for inspectors. It is to find out, and influence, what people are doing when they are not there. There was an undercurrent of violence in the hospital. Troublesome patients were scientifically thumped. Some were illegally restrained by being tied in chairs. In the ward where I worked a 'high grade' patient was set by a window to watch for the doctor so that several minutes warning could be given before his arrival. Admittedly the only fight I saw was between nurses in the corridor outside the hospital office over the disputed contents of a pay slip, and a few patients got their own back by depositing faeces in staff cars and throwing them at you when you walked past their airing court. It also has to be said that some staff were excellent—I will always remember 'Granny' B—s for the example he set. He was given that nickname by his more aggressive colleagues because he fussed over patients and tried to ease their lot.

There is a simple answer to all these problems. It is to close the whole operation down. It has happened for most deep coal mines in the United Kingdom. The hospital for 'idiots and imbeciles' where I worked and the lunatic asylum in Lancaster both shut years ago. The ward block where we cleaned up cohorts of wet and dirty patients before breakfast has been torn down and a development of desirable houses called 'Highgrove' built on the site. Likewise the building where in 1840 Gaskell freed patients from their manacles has been converted into attractive and expensive flats. But unless everybody converts overnight to vegetarianism, and agriculture absolutely eschews the use of manure, this is not a practical control measure for *E. coli* O157. There is an alternative however, which might, in theory, achieve the same aim. The organism is a normal inhabitant of the intestines of cows and sheep. A significant minority of animals carry it. It only causes disease in humans. We catch it by unwittingly eating manure. So eradication of the organism from cattle would get rid of the problem. Unfortunately, at the present time the prospect of doing this is about as realistic as converting everybody overnight to vegetarianism. Nobody knows how to do it. Even if vaccines could be developed it is far from certain whether they would work well enough or whether anyone would be willing to pay for them. We are left, therefore with risk reduction, rather than risk elimination, and must focus on programmes that reduce the amount of manure we eat and that kill the organisms in it.

Chapter 6

E. coli O157

The first recorded human *E. coli* O157 infection occurred in California in 1975, when it was isolated from a woman with bloody diarrhoea and abdominal cramps. This case came to light during a retrospective search for the organism in a collection of more than 3000 *E. coli* strains isolated from patients in the USA between 1973 and 1982 made by the Centers for Disease Control at Atlanta. It was the only *E. coli* O157 in the collection. Similar retrospective searches in Canada (of more than 2000 *E. coli* strains collected between 1978 and 1982) and Britain (of more than 15,000 strains collected between 1978 and 1982) revealed only six cases and one case, respectively. The Centers for Disease Control has no records of outbreaks of bloody diarrhoea of unknown origin occurring before 1982. This makes it very unlikely that *E. coli* O157 had been causing outbreaks but had been missed because laboratories were unaware of it. Bloody diarrhoea is its hallmark. Similarly, in Britain it was not found in any of the 161 diarrhoea outbreaks investigated by the Reference Laboratory of the Public Health Laboratory Service between 1973 and 1983. So the organism is new.

The first *E. coli* O157 outbreak recorded anywhere in the world occurred in February and March 1982 in Medford, Oregon, a medium-sized town located just to the west of the Cascade Range in the Rogue Valley, a fruit-growing area. Many of the 25 patients lived in the suburb of White City. All of them had bloody diarrhoea and abdominal cramps. Some were thought to have appendicitis—one even had a normal appendix removed. Nearly all had eaten at one of two McDonalds restaurants, where 21 had consumed a 'Big Mac'—'two all-beef patties, special sauce, lettuce, cheese, pickles, onions on a sesame seed bun', one had had a regular burger, and two, cheeseburgers. *E. coli* O157 was isolated from four patients. This outbreak made little impact. The full significance of the epidemiological and bacteriological connections that had been made during the investigation was not absolutely clear to the investigators. There was a reluctance to target burgers, or beef, or McDonalds, without stronger evidence. The organism had never caused outbreaks before. Medford might be a 'one-off', a curiosity. But the next outbreak put the organism on the map as the 'burger bug'. It happened at Traverse City, Michigan between 28 May and 27 June 1982. Again, everyone who was ill had bloody diarrhoea and abdominal cramps. Seventeen out of eighteen victims had eaten hamburgers at McDonalds within the previous ten days. *E. coli* O157 was isolated from four of them. The Centers for Communicable Disease Control at Atlanta investigated both outbreaks and presented data about them at a meeting in Miami in early October. In true epidemiological fashion McDonalds was described as 'Chain A', but the press went to work and soon found out its true identity. A story with the headline 'Undercooked Burgers Linked to Disease' appeared in the *Miami Herald*. At 2.45 PM the next day trading of McDonalds Corporation stock on the New York Stock Exchange was halted after it fell one and a half points. But its price soon rebounded. Nobody in the two smallish outbreaks had died or developed life-threatening complications. People continued to eat hamburgers with relish, and with confidence. Nevertheless, the seeds had been sown for the microbiological undermining of trust and faith in this all-American icon. Ten years elapsed before the hammer blow fell. On 12 January 1993 a paediatric gastroenterologist in Washington State reported to the local Department of Health that there was an increase in the number of children coming to the emergency department of the Seattle Children's Hospital and Medical Centre with bloody diarrhoea. In addition, three children had been admitted with the haemolytic uraemic syndrome. Two days later an active search for more cases began, and by 17 January it became clear that a big outbreak was in progress. Of 37 patients 27 had eaten at a Jack-in-the-Box fast food restaurant. A number of different premises were implicated. An announcement was made the next day naming Jack-in-the-Box hamburgers as the probable infection source, and 255,000 patties of the suspect batch of ground beef were recalled. They had all been produced on 19 November 1992 in Los Angeles. At the end of the outbreak 732 people had fallen ill, 195 had been hospitalized, and 55 had developed the haemolytic uraemic syndrome, with four deaths. Three-quarters of those affected were under 18 and the median age was 8.

The first British *E. coli* O157 outbreak was very different. In early July 1985 49 people in East Anglia fell ill, of whom 19 were admitted to hospital in Norwich and Cambridge as medical or surgical emergencies. Bloody diarrhoea and severe abdominal pain dominated again. Three patients had laparotomies, one needed a blood transfusion, and one woman aged 64 died of what was described at the time as 'acute fulminating idiopathic colitis'. But burgers were not involved. Most of the cases had not even eaten out of home in the two weeks before their illness. Three were vegetarians. The most unusual feature of the outbreak was the predominance of women among its victims—38 out of 49. It was concluded that the most likely way that they had become infected was by handling and preparing vegetables, particularly potatoes. A batch had probably been contaminated with manure which contained large numbers of *E. coli* O157 organisms. As often happens in outbreak investigations the ultimate test of the hypothesis—looking for the organism in the suspect food—could not be done because it had all been eaten or thrown away. This is one of the most important reasons why a source is never identified in many outbreaks.

Nevertheless, comparison of those in the US and the UK shows clear national differences regarding the routes of transmission of infection. In the 19 US outbreaks described in the decade after Medford, ground beef was implicated in seven, roast beef in two, water in two, person-to-person spread in three, and vegetables, milk, mayonnaise, school lunches, and apple juice in one each. In the 40 British outbreaks documented in the decade after the one in East Anglia, only two implicated burgers. Two implicated milk, food in general was either definitely implicated or suspected in four, butcher meat in two, water in two, and a delicatessen and turkey roll sandwiches in one each.

Confirmation of the versatility of *E. coli* O157 and its eclectic exploitation of different vehicles for its transmission from manure to mouth in different countries came from Sakai City, Japan, on 13 July 1997. By the end of that day 255 children from 33 elementary schools had received medical attention for diarrhoea and bloody stools. The outbreak went on to grow explosively. By the night of the 14th more than 2000 peple had fallen ill, and the first isolations of *E. coli* O157 had been made from some of them. At the end of the outbreak a total of 12,000 had been affected, two-thirds of them pupils or staff from 47 schools. Fortunately only 121 developed the haemolytic uraemic syndrome. Of these 21 needed kidney dialysis and three died. Investigations focused on school meals. Because each school cooked its own food, particular attention was paid to ready-to-eat foods. Bread and milk were excluded, leaving only one common source— *kaiware daikon* (radish sprouts), epidemiologically identifying this dish as the villain of the piece. It so remains, although *E. coli* O157 was never isolated from it directly—or found during the exhaustive investigation of other foods; 985 samples from 362 sources were tested.

  E. coli O157 is feared because it kills. It is a member of the traditional class of plagues, like tuberculosis, smallpox, and plague itself, and is old-fashioned, like these diseases used to be, and smallpox still is, in that there is no specific curative remedy. Another of its unpleasant attributes is that even when it does not kill, it leaves some of its victims damaged for life. Kidney function can be permanently impaired or even completely destroyed. While dialysis and kidney transplants can preserve a tolerable quality of life, the irreversible brain damage that it sometimes causes cannot be alleviated technically and can leave individuals totally dependent on their carers. These complications are formally called the haemolytic uraemic syndrome (HUS), and thrombotic thrombocytopaenic purpura (TTP). In both of them the main symptoms and the damage to organs are caused by the death of cells that line small blood vessels, in the kidneys and brain. There are changes in the cells that circulate in the blood stream as well. These effects are produced by a family of closely related poisons produced by the bacterium, called Shiga toxins (Stx). *E. coli* O157 does not invade the blood stream or tissues but causes all its mischief while sitting on the cells that line the intestine. The Stxs stick loosely to the surface of white cells in the blood and travel on them to their targets in the kidneys and brain. They also damage the intestines directly. For the organism to do these things it has to live in the intestines for while. It uses a very sophisticated mechanism to achieve this. It produces a protein called *tir*, which it exports. *Tir* is taken up by enterocytes, cells on the outside of microvilli, the microscopic finger-like projections that line the gut, and arranges itself so that its molecules jut out of the cells. Intimin, a protein on the outside of the bacterium, recognizes these molecules, and sticks to them. Various changes in the microvilli then occur, and their relationship with the bacteria becomes very intimate, the microvilli changing shape from a finger to a cup. The *E. coli* O157 ends up sitting in the hollow.

Only a minority of individuals infected with *E. coli* O157 develop HUS or TTP. A substantial proportion have no symptoms at all. It is impossible to predict what will happen when someone is exposed to infection. All that can be said is that age is very important, HUS and TTP being much commoner in children, particularly those aged 5 and below, and in old people, as was so evident in Wishaw. They are much rarer in 20- to 50-year-old people.

Why does *E. coli* O157 cause such large and dramatic outbreaks? One that occurred in North East Scotland just two weeks after Walkerton is particularly instructive. As part of the Millennium celebrations Scouts throughout the UK held summer camps. North East Aberdeenshire planned to hold theirs between Friday, 26 to Sunday 28 May at the New Deer Agricultural Show Ground. Their arrangements were frustrated by the weather, however, and the camp was abandoned a day early because of the heavy rain which caused part of the showground to become waterlogged. *E. coli* O157 went to work. Of the 20 campers from whom the organism was isolated (another 50 of the 337 cubs, scouts, venture scouts, leaders, and helpers at the camp had gastrointestinal symptoms) the first two fell ill with diarrhoea on Sunday, 28 May. Two more came down the next day, seven on the 30th, four on the 31st, three on 1 June, and one on the 2nd and one on the 3rd. One had kidney problems and needed dialysis to tide him over until their functions returned properly. Food and water were confidently ruled out as sources and suspicion fell on the 300 sheep that had been grazing the showground until the day before the camp. They had been brought in to shorten the grass and had left behind an abundance of faeces containing *E. coli* O157. Questioning the campers and some simple calculations showed that those who did not wash their hands before meals were nine times more likely to be ill than those who did, and that those who did not use cutlery for meals were seven times more likely to be ill than those who used knives and forks. My Aberdeen colleagues Norval Strachan, David Fenlon, and Iain Ogden took things further by working out how many *E. coli* O157 the scouts had eaten. They started by finding out the number of organisms the sheep had dropped on to the campsite. Considering that the average daily mass of faeces produced by a sheep is 1 kg, knowing the number of *E. coli* O157 per unit mass of faeces for a New Deer lamb and a New Deer ewe, and taking into account the build up of organisms over the six days that the sheep were grazing before the camp, the decay of the organisms over time in old faeces, the number of sheep and the area of the campsite, they calculated that a gram of soil from the field on the days of the camp contained 60 *E. coli* O157 (Figure 6.1). This figure was very close to the number found in actual samples. Feeding

Figure 6.1   Turds on the campsite track at New Deer: the 'smoking guns'.

into their calculation an estimate by Dutch researchers that children on camp sites eat between 30 and 200 mg of soil every day showed that the infected scouts had ingested between 4 and 24 organisms. So the infectious dose of the organism—the number needed to start an infection—was extremely small. This result is very similar to those of studies on leftovers from food that caused outbreaks. New Deer also confirmed something else that was known before—that *E. coli* O157 cannot only be carried and excreted by farm animals like cattle and sheep, but that these infected animals are perfectly fit and healthy. Without laboratory tests there is no way that a farmer or even a veterinarian can tell whether an animal is a carrier. *E. coli* O157 carriage in farm animals is not routinely looked for. Two big research surveys have shown that in England and in Scotland between 4.7 per cent and 8 per cent of individual cattle carry it and that 23 per cent to 44 per cent of herds are positive.

Follow up studies of the New Deer camp site showed that the organism was surprisingly persistent. It did not disappear from soil samples until January 2001; one of two taken in December 2000 was positive.

The commonness of carriage of *E. coli* O157 in farm animals, its low infectious dose for humans, and ability to survive for a long time in the environment help to explain why it causes big outbreaks. A little uncooked manure can clearly go a long way. The really surprising thing is not that big outbreaks occur, but that they are so rare. After all, bad hygiene practices had been going on in Barr's for years, the Walkerton water works had been cutting corners for more than a decade and using the high risk Well 5 for 20 years, and Americans have been asking for rare burgers ever since they were invented. Work on the New Deer sheep after the outbreak provides a possible explanation. While the average number of *E. coli* O157 per gram of faeces being shed by ewes was 8400, one lamb was shedding more than a million. Such animals are almost certainly uncommon. But they are very dangerous. It may be that their relative rarity explains the infrequency of outbreaks. The large numbers of bacteria they produce would clearly help to counteract the dilution effect of faeces dissolving in a town water supply, being mixed into a batch of ground beef prepared from many animals, or being transmitted through butchery premises by cross-contamination.

Chapter 7

Other *E. coli*

Large and dramatic *E. coli* O157 outbreaks started to occur only in the 1990s, presumably because that was when the organism first firmly established itself as a member of the normal bacterial ecosystem of cattle and sheep intestines. Where had it come from? We will never know for certain, although the prevalence of the haemolytic uraemic syndrome in middle-class Argentinian children from the 1960s onwards—admittedly without proof that they followed an *E. coli* O157 infection—hints tentatively at an origin in cattle on the pampas. It is well known that babies in that country and that social-economic group are weaned onto beef. One can be a little more definite about the evolutionary time scale of the origin of *E. coli* itself. Studies on gene sequences indicate that it began to evolve as a separate species from its close relative *Salmonella* about 140 million years ago. *E. coli* was first described by the paediatrician Theodor Escherich in 1885. In the first detailed study of a bacterium that infected humans but was not thought to cause a specific disease, he investigated the faeces of new born infants and showed that it appeared in large numbers soon after breast feeding started. He considered it to be harmless, and in general, he has turned out to be correct. Most individual *E. coli* bacteria only cause problems if they escape from the intestines into the peritoneum, or move up the urinary tract and infect the bladder, and, sometimes, the kidneys. Their presence in the intestines, however, passes without notice as far as the host is concerned. Its world population at any one time is about 100 billion billion cells, nearly all living in the intestines of people and animals, with about 200 generations a year.

E. coli bacteria reproduces without needing sex. So all the progeny of an individual are clones of it. But over the millions of years of its evolution some clones have become commoner than their sisters. For some this has been through chance. For others, sex, or something like it called parasex, has been important. *E. coli* bacteria can exchange DNA and can also receive DNA when they are infected with viruses. They also can reshuffle their own DNA to make the best use of it by rearranging their genes. Mutations happen all the time as well. All these genetic changes can be advantageous. It has been calculated that a clone with a fitness (competitive advantage) of 1.00001 could increase to a population of 10 billion billion cells in 22,000 years. On an evolutionary time scale this is very very rapid. What has happened in practice is that a large number of clones has evolved. Many are rare but a few are quite common. Most are harmless when in the intestines. But a few, like *E. coli* O157, cause disease.

Notable outbreaks in Britain during the last century give useful information about the evolution of *E. coli* clones, and the diseases caused by them.

E. coli O157 has been described as a developing world infection occurring in the developed world. Such a developing world disease is bacillary dysentery, which caused dramatic outbreaks. It killed 8300 in Guatemala in 1968-9 and 15,000 in the Rwandan refugee camp at Goma, Zaire in 1994. In the developing world as a whole it is currently estimated to cause about 200 million cases and kill 650,000 people every year. The causative organism was first described in 1898 in Tokyo by the bacteriologist Kyoshi Shiga. It was later called *Shigella* in his honour. There are several different kinds. Shiga's bacillus is now called *Shigella dysenteriae* type 1. It produces toxins that are closely related to those of *E. coli* O157, the reason why they are called Shiga toxins. Like *E. coli* O157 the Shiga bacillus sometimes produces HUS in its victims. Other shigellas do not have these toxins. *Shigella flexneri* is the commonest worldwide; *Shigella sonnei*, the mildest, has dominated in Europe and North America since the Second World War. By the end of the nineteenth century, as a cause of death in British adults, *Shigella* had retreated to a single fastness, the lunatic asylum. We have seen from the records of Stanley Royd that it had been in the hospital from the beginning, not always causing a lot of cases, but lurking there, taking off from time to time to cause outbreaks.

Early bacteriologists often had difficulty in distinguishing the bacteria they isolated from cases of dysentery from *E. coli*. With hindsight this is easy to explain, because we now know that *Shigella* strains are nothing more than *E. coli* clones. They may behave differently from run-of-the-mill *E. coli*, but genetically they are all members of the species. The essential difference between them is that *Shigella* has acquired a portfolio of genes on a small piece of DNA which code for proteins that it uses to invade cells in the large bowel. At the beginning of an infection the bacterium is taken up by cells in the intestine that belong to the immune system. It subverts the machinery inside these cells and uses it to move to locations from which it then invades the intestinal cells themselves. All this leads to much inflammation, which initially causes a lot of damage. This is why there is often blood in the diarrhoea and why severely affected patients develop bowel ulcers. In the majority of previously healthy victims, however, the inflammation gets rid of the bacteria and cures the infection.

Asylum dysentery was notable for two things. First, how it persisted in these institutions despite getting rarer and rarer in the community, and secondly, its remarkably high mortality. An infection in an asylum was about 100 times more likely to kill than one in the community. At the end of the nineteenth century its impact on asylum life and of pressure from the Lunacy Commissioners caused some young asylum doctors to turn into bacteriologists and infectious disease specialists in order to sort out the problem. High-quality work was done. A detailed study done at Lancaster by J. F. Gemmel was typical. In his book on dysentery published in 1898 he said that

> a more than unusually severe outbreak of what is variously termed ulcerative colitis or dysentery, having occurred in this asylum in the early part of 1895, and the appearance of a similar, though milder one towards the end of autumn, have induced me (to describe and analyse) clinical data of more than ten years of this most distressing and fatal affection.

One of the typical cases he described was that of Susan C—, aged 75, a 'senile maniac, resident for 3 years'.

> August 2nd, 1895. Yesterday, the patient, a fairly vigorous female for her age, complained of feeling cold and shivery and had an attack of bilious vomiting, followed later by scanty mucous and bloody purging and pain in the abdomen. Temperature 102.6°, fell to 101.6° in the morning. Treated with milk, beef soup and stimulants, and half an ounce of cream of tartar daily. August 4th. Purging still continues. August 6th. Diarrhoea not so severe, but patient's general condition grave; voice

almost inaudible, hollow cheeks, sunken eyes and feeble pulse and coldness of extremities. August 8th. Bowels move about four times in 24 hours, evacuations scanty and very offensive, herpetic eruption on lips, tongue dry and brown. Temperature normal. Saline discontinued but stimulant freely exhibited. August 10th. Temperature last night 101°, fell this morning to 99°. Patient has been purged six times since yesterday morning and is sinking fast. She died in the course of the day. Post mortem. Enlarged mesenteric glands, Caecum slightly congested. Descending colon mucous membrane red, rough and granular looking. In lower part of sigmoid there are large solitary ulcers, their edges undermined.

Many of those who studied asylum dysentery at this time concluded, like Gemmel, that regarding 'its specific course, we have strong evidence to show that it is dependent on a specific microorganism'. But there was another school of thought which claimed that in certain lunatics the vitality and resisting power of tissues to infection was reduced owing to the impairment of their 'trophic' nerve supply, in particular that of the colon, allowing normal and harmless intestinal bacteria to grow unchecked and cause disease. 'These intestinal lesions form a part of the general degenerative process. They owe their origin to a nervous perversion.' This was asylum ear all over again. But at the end of the day Gemmel's views prevailed. A report in 1900 by Frederick Mott and Herbert Durham on 'colitis' to the London County Council, which ran many asylums, was influential. It caused the Lunacy Commissioners to issue a circular about it, following which the number of cases was recorded and much attention was paid to control measures. Mott's view was the 'much asylum dysentery is due to communication of the disease from one patient to another, probably by the ignorance, carelessness, or deficient precautions on the part of the attendants'. Gemmel presciently hypothesized that poor diets predisposed patients to a serious outcome, and said that 'perusal of asylum reports will show (it) is often associated with overcrowding'. The correctness of these views was dramatically shown during the First World War. Asylums became more overcrowded because about 10 per cent of them were turned over to the Army for military use. Many attendants enlisted, so staffing levels fell, and there were severe reductions in the diet of patients. In consequence there was an enormous rise in the number of deaths from dysentery. More than a thousand patients died from it in 1917.

Asylum dysentery came under control when measures that prevented the faeces of infected persons getting into the mouths of others—often via the hands—were effectively implemented. Incontinent patients were a particular problem, of course. Dysentery is no longer a problem for institutions looking after such patients. *E. coli* O157 has replaced it. Hartwoodhill Hospital (Chapter 1) was a classic example.

Nobody now believes that diseases like schizophrenia or depression impair the 'trophic' nerve supply to the intestines. The very concept of such 'trophic' nerves has disappeared. But the notion that there was a relationship between these diseases and bacteria did not go away easily. First, it was turned on its head. A school of early twentieth-century enthusiasts came to believe that chronic infections caused them. The 'detoxication' regime of Henry Cotton, Medical Director of the New Jersey State Hospital at Trenton, was particularly vigorous (Figure 7.1). On the basis of finding that various streptococci and *E. coli* (of the sort that we know now to be harmless and part of our normal bacterial flora) were common in psychotic patients, he decided in 1918 in his own words to 'clean up' his patients of 'all foci of focal sepsis'. Teeth were removed whenever there was the slightest suspicion of abnormality—up to five extractions from every mouth. Sockets were scraped out. Tonsils were removed from about 90 per cent of patients. In women the cervix was next to go, and sometimes the uterus, tubes, and ovaries. The seminal vesicles were removed in about half of chronic psychotic men. When patients showed no improvement, 'the colon then becomes the organ to be eliminated'. In his series of 133 total colon removals, 33 recovered mentally, but 44 were killed by the operation. Partial resection patients fared no better, with 44 recovering and 59 dying out of a group of 148. British practice in the 1920s was less drastic. In the Edinburgh Royal Infirmary, for example, the bowels were addressed by purging followed by intestinal 'wash outs' with warm water, the administration of live yoghurt, and massage and remedial exercises to improve 'the vascular supply and nerve tone of the structures which regulate the action of the liver, adrenal glands, and other abdominal viscera'. 'Detox' of this kind has, of course, survived into the twenty-first century. But in the century since it was invented, its evidence base has made no progress. It is as utterly feeble now as it ever was. The persistence and popularity of such pseudoscientific quackery and the willingness of people to pay for it provides eloquent testimony to human gullibility and our capacity for self-deception.

Many people use the word 'natural' to denote things that are good and wholesome. The philosopher John Stuart Mill wrote an essay to test this notion, 'to inquire into the truth of the doctrines which make Nature a test of right and wrong, good and evil, or which in any mode or degree attach merit to approval to following, imitating or obeying Nature'. He concluded that for people to follow Nature—to make the spontaneous course of things the model for their voluntary actions—is irrational and immoral: 'Irrational because all human action whatever, consists in altering, and all useful action in improving, the spontaneous course of nature: immoral, because the course of natural phenomena (is) replete with everything which when committed by

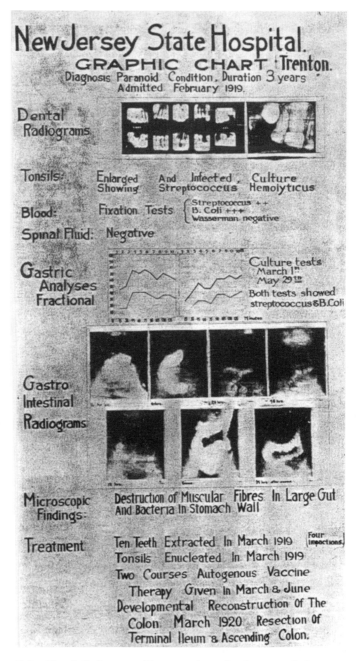

Figure 7.1 The full detox at Trenton: 'treating' paranoia by removing teeth, tonsils, and bowels.

human beings is most worthy of abhorrence'. To quote two of the
many ways in which he illustrates Nature's callous indifference:
'Anarchy and the Reign of Terror are overmatched in injustice,
ruin, and death, by a hurricane and a pestilence'. One of the most
'natural', and nastiest, features of human life has, until very recently
been the killing of babies by infection. It took a long time for bacteriologists to work out the causes of the diarrhoea that was so lethal for
them. Just as with asylum dysentery, in many cases it was because
particular *E. coli* clones were responsible—clones which for technical
reasons were difficult to separate from the harmless ones discovered by
Escherich and which predominated in the intestines. A landmark event
in the study of these organisms was the Aberdeen *E. coli* outbreak of
1947, because it gave John Smith the opportunity to do this. A billiards-
playing orchid-growing Aberdeen bacteriologist, he had already won
prizes for his pioneering researches on Streptococci in childbed fever.
Although Scotland had a poor record in child health with relatively
large numbers of deaths in the first year of life—the period used to
measure the infant mortality rate—this had been falling steadily since
the First World War. In Aberdeen it had dropped from 140 per 1000
live births in 1915 to 50 in 1946. But in 1947 it rose to 64—almost
entirely due to an epidemic of gastroenteritis. Smith worked out his
own system for classifying different kinds of *E. coli* and found a
particular strain that he called alpha in 219 sick babies; 99 of them died.
Nearly 90 per cent of those who fell ill were in the first seven months
of life, and 98.1 per cent were being bottle fed. Just over half of them
were already in hospital when they contracted their infections, and so
it was clear that the bacterium's nastiness was significantly helped if its
victims were already enfeebled by other health problems. But an
outbreak among the new-born in the maternity hospital, showed the
significance of age alone as a risk factor. In 1948 there were many fewer
E. coli alpha cases—25 with six deaths—but another type, beta,
appeared, infecting 37 and killing 11. The alpha and beta strains eventually disappeared from Aberdeen and there were no more cases after
the middle 1950s. Comparison of the Aberdeen types with the international identification system that was developed soon after Smith's
work showed that alpha was *E. coli* O111, and beta was *E. coli* O55. They
are now called enteropathogenic *E. coli*. Why these organisms were so
common in Aberdeen in the 1940s has never been explained, nor is it
understood why they went away. There have been few outbreaks in
Britain in recent times. But the ability of these clones to get about
geographically was convincingly shown by a particular one which was
first identified in Indonesia in 1960. In 1967 it was found in Dundee
and in 1969 in Glasgow, where it continued to cause hospital outbreaks

until 1971. In 1970 it appeared in a hospital in Stoke-on-Trent and in 1971 it caused problems in Leeds, Manchester, and Liverpool, as well as in Belfast and Dublin. It spread to Canada in 1972 and Arizona in 1975. Not a lot has been seen of it since. Enteropathogenic *E. coli* have the same machinery for sticking to intestinal cells as *E. coli* O157, but they do not have any shiga toxin genes. It has even been proposed that *E. coli* O157 is descended from an *E. coli* O55 clone. All it needed to do was acquire the toxin genes—in molecular genetic terms a relatively straightforward and very believable step to have been taken.

E. coli does not have a skeleton and so fossils of it are not available to help in studies of its evolution. Nevertheless, this has not stopped us from constructing very plausible family trees of its ancestry going back millions of years. How is this possible? The task is no more difficult than the one faced by the fossil hunter if instead of studying its bones one looks at how closely related its genes are to those of other bacteria, how they use the genetic code to make for proteins and other molecules, and how its genes are arranged. *E. coli* has a single chromosome made of one long DNA molecule joined at its ends to form a circle. It is usually called the genome. Its complete nucleotide (base) sequence—the precise arrangement of its building blocks, each nucleotide being a letter in the genetic code—was first determined in 1997. An *E. coli* strain called K12 was picked for this analysis. It was isolated from the faeces of a convalescent diphtheria patient in 1922 and for many years had been used in the practical classes run by the bacteriology department at Stanford University in California. The geneticists Edward Tatum and Joshua Lederberg had used it in the 1940s to show that bacteria had sex. It was a lucky choice for them because later comparison showed that for an *E. coli* it was particularly virile and fertile. Tatum and Lederberg were the first scientists to win a Nobel Prize for work on *E. coli*. They were not the last.

In 2001 the genome sequence of the 1982 Traverse City outbreak strain of *E. coli* O157 was published. Comparison with the K12 sequence confirmed what was already known—that O157 had genes not present in other *E. coli* strains—but it was full of surprises as well. Both had a common sequence backbone of about 4.1 million bases. With the exception of a chunk of 422 thousand bases which ran in different directions in the two genomes, they lined up well. But the basic backbone was interspersed with hundred of 'islands' of DNA, up to 88 thousand bases long, found in one strain but not the other. O157 had 177 unique islands longer than 50 bases and K12 had 234. About a quarter of the O157 genes, 1387 out of 5416, were within the 'island' DNA. The function of only about 40 per cent of them could be worked out from the similarity of their sequence to known genes.

The completion of the *E. coli* O157 genome sequence was eagerly awaited. There was a naive hope that somehow it would lead to a breakthrough in our understanding of why the bacterium is so nasty, and where it had come from. But its main impact has been to remind us how much more work has to be done before we can have even partially complete answers to these questions. It has identified, for example, lots of genes whose function is completely unknown. It has also whetted our appetite for more genome sequences, so that the genes needed for virulence—the ones that distinguish disease-causing strains from harmless ones—can be identified with certainty, and so that the relationship of O157—enterohaemorrhagic—strains to other virulent *E. coli* like the enteropathogenic infant enteritis strains can be worked out.

None of these studies would be possible without the methods and approaches of molecular biology. There is an enormous paradox here. It is that despite our current levels of ignorance about it, which is very great, *E. coli* itself played an enormously important role in laying the foundations of molecular biology itself. One crucially important factor was the nature of the organism and the techniques that had been worked out to handle it. The bacterium grew so rapidly that the results of experiments on it were available in less than 24 h; breeding experiments could be done hundreds of times faster than with Gregor Mendel's peas, for example. Millions of organisms could be handled on a small dish in the laboratory, and mutants could easily be identified using methods initially worked out to separate virulent bacteria like those causing typhoid from its harmless relatives. Incorporating a simple molecule, lactose, in the agar jelly on which the bacteria grew and a litmus-like indicator molecule that changed colour when acid was produced enabled overnight growths of clusters of bacteria, like typhoid, that left lactose unchanged, to be distinguished from the other intestinal bacteria that metabolized it with acid production. The colonies of typhoid bacteria remained white. All the others went red. These methods were very cheap. All they needed were some reusable glass dishes with lids, some culture medium made of simple ingredients like meat extracts, an incubator, an autoclave (a pressure cooker) for sterilization, a wire loop or some sterile orange sticks to pick up interesting colonies, and a bunsen burner.

Also crucial were the viruses that infected *E. coli*. Called bacteriophages, or phage, they were easy to detect. Growing a 'lawn' of bacteria on the surface of an agar plate—where the culture medium had been solidified—infecting them with phage and incubating overnight led to the development of clear holes, or plaques, in the bacterial lawn. Building up a collection of phages for experimental work was easy. Town drains are a rich source and many of the ones that are used today came originally from the Los Angeles sewage system.

All these properties of *E. coli* and its viruses were necessary for rapid experimental progress. But they were far from sufficient. It was not until they came to the attention of a remarkable group of scientists in the 1940s that everything fell into place. Since then more Nobel Prizes have been awarded for work on *E. coli* than any other species except the human. In the 1930s it was only possible to speculate about the answers to three questions—what are genes made of, how do they reproduce, and how do they act. Using *E. coli* and its phages, scientists in the United States in the early 1940s like Max Delbrück, Salvador Luria, and Alfred Hershey, began work which answered them. They have been described as the three bishops of the 'phage church'. Delbrück was the pope. He had trained in Germany as a quantum physicist but had been influenced by Niels Bohr, the Danish physicist and Nobel Laureate, to consider that life might not be reducible to atomic physics. He went to the United States in 1937 to work on genetics to test this idea. Luria had been a doctor in Italy but had also become a geneticist. Hershey was the only American of the three. His reticence was famous. It was rumoured that when he was particularly garrulous he once said 'please pass the pipette can'. The work that these scientists did on *E. coli* was enormously influential. It drove Watson and Crick's obsession to find the structure of DNA. *E. coli* was not only particularly suitable for geneticists but was an excellent subject for biochemistry. The Hershey–Chase experiment exemplifies this brilliantly. It is one of only a handful of experiments in biology that is referred to by the names of those who did it. It was enormously influential in convincing scientists that genes were made of DNA. Hershey, and Martha Chase, his research assistant, labelled phage with radioactive isotopes—sulphur, which only went into protein, and phosphorous, which went entirely into DNA. They infected *E. coli* with the 'hot' virus, and stripped off the phage that was left sticking to the bacterial surface with a Waring blender, a 'whirling blades' food processor. Much more DNA than protein entered the cells, and little of the protein was incorporated into new virus particles. In true cautious Hershey style (Figure 7.2) the conclusion was drawn that

> the sulfur-containing protein of resting phage particles is confined to a protective coat that is responsible for the absorption to bacteria, and functions as an instrument for the injection of the phage DNA into the cell. This protein probably has no function in the growth of intracellular phage. The DNA has some function. Further chemical inferences should not be drawn from the experiments presented.

While Delbrück, Luria, and Hershey were busy in the United States, others were also exploiting *E. coli*. In Paris Jacques Monod had picked

Figure 7.2   Leo Szilard (left), about to speak to Alfred Hershey, 1951.

it for his work on growth after a senior colleague at the Pasteur Institute, André Lwoff, had replied 'No' to his question 'Is it pathogenic?' During the day he did his *E. coli* research on genes and how they worked at the Sorbonne. For the rest of the time he was active in the resistance against the Germans. He became executive officer to the chief of staff of the Francs-Tireurs et Partisans, run by the Communist Party. Eventually, under the name 'Malivert', he became chief of staff for operations for all resistance forces in France. He had had to leave the Sorbonne because his identity was becoming known—he was briefly arrested by the Gestapo in 1940—but was given bench space by Lwoff at the Pasteur Institute. Monod, Lwoff, and another French colleague, François Jacob, were awarded the Nobel Prize in 1965 'for their discoveries concerning the genetic control of enzyme and virus synthesis'. Delbrück, Luria, and Hershey followed in 1969 'for their work with bacteriophages'.

Like Tatum and Lederberg before them, none of these scientists worked on *E. coli* because of its medical importance. This was not because they lacked an understanding of medical matters. Lederberg had started a medical degree but switched to genetics. Luria had an MD from the University of Turin. Lwoff had obtained a medical degree to please his father, who was a doctor in charge of a lunatic asylum. In 1940 Jacob was a 2nd-year medical student. He fled France to join deGaulle's Free French Forces and was wounded in Tunisia and in Normandy in August 1944. He qualified, worked on the French penicillin programme, got bored, and became a researcher. The tradition exemplified by these scientists of using *E. coli* not as a medically important bacterium but as a model system to answer big biological questions was justified by Monod when he said that 'anything that is true of *E. coli* must be true of elephants, except more so'. It has also spawned enormously important practical applications. Werner Arber won the Nobel Prize in 1978 for his work on *E. coli* restriction enzymes. These are enzymes that defend bacteria against invading DNA by cutting it into fragments. Genetic engineers use them when they stitch pieces of DNA together to construct products like vaccines and genetically modified plants. Arber shared the prize with Daniel Nathans, a virologist, and Hamilton O. Smith, another doctor turned molecular geneticist.

So we know an enormous amount about the molecular biology of *E. coli* but at the same time are really quite ignorant about many aspects of its natural history in the real world. The explanation for this is simple. It is technically easier (and a lot cheaper) to probe its structures and functions in the laboratory than to follow its spread through human populations or herds of cattle, or to find out precisely what it does in the body of a child. But at least with *E. coli* a deep scientific understanding of the bacterium gives us a firm basis and a solid foundation on which to remedy our ignorance about its behaviour in nature. The same cannot be said for BSE and CJD.

Chapter 8

CJD

Stephen (Steve) Churchill was born on 14 April 1976 in Stockton-on-Tees. He was a second child. His father was a fireman, and the family lived in a large Victorian semi-detached house. Steve was a lively child. He played rugby and was a good swimmer. When he was 12 his father obtained a promoted post in the Wiltshire Fire Brigade and the family moved to Devizes. Steve gave up rugby because his lanky physique was not right—he was growing tall but thin—and he joined the Air Training Corps (ATC). His big ambition was to fly with the RAF when he left school. But in the summer of 1994 things began to go wrong. He did badly in school examinations. He resigned from the ATC. In August he wrecked his mother's car by driving on the wrong side of the road and colliding with an Army lorry. In September the results of his school re-sit examinations were not good. In October and November his feelings of inadequate performance intensified, and he left school. His general practitioner diagnosed depression, and a consultant psychiatrist made the same diagnosis in December. Anti-depressants were prescribed. Up to Christmas he spent most of the time at home watching television. He began to have panic attacks. A scene from the children's drama 'Black Beauty', where a little girl was hiding from someone in a hay loft, made him agitated and fearful as though he was the person who was hiding from danger. He had hallucinations, ate less and less, began to have pains in his side and in his leg, and lost the ability to sign his name. Christmas was miserable. In previous years he had enjoyed it as the highlight of the year, but in 1994 he fell asleep during the dinner and left many of his presents unopened. On the 3 January he was admitted to a psychiatric hospital. His physical and mental decline continued. He lost weight and developed a staggering gait; his panic attacks—'daymares'—got worse and sometimes he injured himself. Electroconvulsive therapy was suggested but rejected by his parents. By now, although a diagnosis of a deep depression was still being held, thoughts of a neurological disease were becoming stronger in the minds of his doctors. His condition deteriorated suddenly in February and he was transferred to the care of a neurologist at a district general hospital 20 miles from the Churchill home. Many tests were done. The ward was understaffed and there was a problem with feeding because of it. His parents decided that they had to visit daily and stay for hours so that he got looked after properly. By March he was totally dependent on others for toileting, washing, and feeding. He was transferred to a specialist hospital in London for a brain biopsy. He was returned to the local hospital on 21 March in the same state as he left. He lived another six weeks. After administrative struggles for funding, a transfer to an old folks nursing home was arranged on 4 May. He died on the evening of 21 May.

The summer of 1994 was also the time when things also began to go wrong for Gulcan Hassan, a 16-year-old Londoner. Her problems had started in March when she complained of a painful left foot. There was nothing to see on examination and physiotherapy did not help. The pain spread to her back and her right knee and by August her walking was affected. She had trouble keeping her balance and her speech was becoming slurred. Intensive neurological investigations were done but a lumbar puncture, CT and MRI scans, bone marrow aspiration, and electroencephalography revealed nothing abnormal. She started to complain of numbness in her face and hands and had a constant urge to urinate. She cried incessantly because of the pain, became wheelchair-bound, and could only eat pureed food. For many months this went on without cessation or improvement. A repeat of the intensive battery of neurological tests in July 1995 again gave negative results, and it was decided to do a brain biopsy. It was diagnostic of Creutzfeld-Jakob Disease (CJD). Within two weeks of the biopsy operation Gulcan had a fit. She had developed a brain abscess, which was treated with antibiotics. Her fits continued, however, and she needed anti-convulsant drugs to control them. By November 1995 feeding was being done by naso-gastric tube, and in December she became bed fast. She lived until May 1997. After two heart attacks and a very short spell in a coronary care unit on life support, she died on 23 May. Her father said in his statement to the BSE Inquiry 'my only ray of consolation is that after her death, Gulcan looked peaceful and beautiful. For the first time in years, her face was smiling. She was in no more pain. Her suffering had ceased.'

It is plain that Steve and Gulcan's illnesses were singularly unpleasant. They both had insight in the early stages that something really bad was happening to them—Steve knew that 'he was going nutty' and Gulcan said that she didn't want to die. Later, in both of them it seemed that the disease disinhibited or even stimulated the part of the brain responsible for feelings of fear and apprehension. They died after a long period of inexorable mental and physical decline. The inability of doctors to make a diagnosis added to the distress of their parents, and brain biopsy brought no relief either through treatment or by a more accurate prediction of the future course of the illness. But with the benefit of hindsight it is possible to understand the diagnostic difficulties, and even why such a desperate and dangerous test was resorted to. It was because a new disease had emerged. Its appearance was not entirely unexpected. Perhaps the worst fears of those who had considered whether BSE could infect humans had been realized.

Four years before Gulcan and Steve's illnesses began a CJD Surveillance Unit had been set up in Edinburgh as one of the official responses to the emergence of BSE. It started work on 1 May 1990 with funding from the Department of Health (England and Wales) (90 per cent) and the Scottish Office Home and Health Department (10 per cent). It was run by Dr Bob Will, a consultant neurologist with extensive experience of CJD surveillance work. Its remit was twofold, to look for changes in the epidemiology of CJD, and to assess the extent to which any such changes were linked to the occurrence of BSE. These tasks were not going to be easy. Identifying the number of particular medical conditions in the population might seem to be a straightforward task. It is not, even for well-defined ones. Formal systems of data collection like statutory notification by general practitioners and specialists and death certificates filled in by them are notoriously inaccurate. This is partly because diagnosis is itself error-prone—port-mortems often show conditions that were missed in life, for example—and because doctors vary greatly in their keenness to fill in forms. The problems for the CJD unit were greatly compounded in that it was not possible to define with any precision what they should look for. If BSE was infectious for people, what sort of disease would it produce? It was already known that human diseases in the CJD family took several forms. In some dementia dominated. In others it was problems with walking and movement. Human BSE might be like one or the other, or it could be completely different. Even after death, diagnosis was not necessarily going to be easy, because the pathology of CJD—essentially the appearance of brain tissues under the microscope—is far from straightforward. The best example of how complex and difficult this is that it took nearly 50 years of study before the spongy appearance of the brain came to be regarded as a crucially important microscopic feature of the disease. The CJD expert Paul Brown has elegantly explained this difficulty:

> nothing is more difficult to evaluate than a space, and nothing is more readily and reasonably attributed to postmortem fixation artefact. Furthermore, spongiosity of brain tissue may also occur in association with premortem states of cerebral hypoxia, cerebral oedema, circulatory disturbances, toxic encephalopathies, and metabolic disorders.

So the CJD unit faced several potentially severe scientific and clinical difficulties. Deliberately setting out to look for an undescribed and, maybe, non-existent disease was a task without precedent. It also had to address these problems under the intense gaze of many interested parties. Nearly every unusual case of CJD—suspect or confirmed—was already being followed up aggressively by journalists. Government ministers received regular briefings, particularly about cases in farmers. Will also regularly reported his findings to the Spongiform Encephalopathy Advisory Committee, of which he was a member. The work of his Unit was also reviewed in its early years by a Medical Research Council Committee chaired by Ingrid Allen, a neuropathologist from Belfast, a rigorous scientist and a person falling firmly, it could be fairly said, into the category of 'grand dame', a category later confirmed officially through the honours system.

Steve Churchill and Gulcan Hassan were the first cases of variant CJD to be referred to the Unit, in May and August 1995 respectively. They were so unusual and so young that they were described in two letters to *The Lancet* in October. Only four previous cases of CJD in teenagers had occurred anywhere in the world, a 16-year-old boy in the United States in 1976, a 19-year-old girl in France in 1982, a 14-year-old girl born in England but resident in Canada, in 1988, and a Polish girl of 19 in 1991. But four more possible CJD cases in people under 50 were referred to the Unit in the July, August, and September, and another seven before the end of the year. Eight, including Steve and Gulcan, turned out to have variant CJD. It took the Unit until

March 1996 to gather enough evidence to make sure that what they were seeing was a new disease, and that there was a particularly British component to it—presumptive evidence of a possible link with BSE.

The Unit did this in a systematic way. It set criteria for novelty, and for establishing the possible extent of linkage to BSE. As tests of the first, assessments were done of the clinical phenotype (what the patients complained of and what was found when they were examined, including the results of clinical tests), the neuropathological phenotype (the appearance of the brain and other tissues under the microscope), and whether the CJD could be explained by particular mutations (established by DNA tests). To look for links with BSE, it was asked whether the UK cases were different from previous experience, and whether they were only occurring in the United Kingdom. As data came in during the early part of 1996 the Unit became more and more convinced that it was seeing something new. By early March it was certain enough to justify publication. The final step was to check with European colleagues whether their surveillance had turned up anything similar. They had not. On 16 March an emergency meeting of SEAC was held to discuss the Unit's conclusions. The Committee agreed that the new variant of CJD was a distinct entity, and concluded that the possibility of a link to BSE had to be taken very seriously.

Although the Unit did first-class work in identifying variant CJD it was not working in a scientific vacuum. For many years Creutzfeld-Jakob disease had attracted much more attention than its rarity would dictate. One explanation for this is that it is a fundamental human attribute to be much more interested in things that are uncommon and dramatic rather than those that are common and mundane. This is the Lardner effect again; although Creutzfeld-Jakob disease was vanishingly rare throughout the twentieth century, its association with cannibalism gave it a cachet unmatched by any other brain disease. Kuru first became common among the Foré people in the eastern highlands of Papua New Guinea during the Second World War after a big high-mortality outbreak of dysentery. The illness was commonest in women. It also affected children. After starting with headaches and limb pains, tremors developed and difficulties with walking and speaking became marked. Dementia was uncommon. Death came after 12 to 18 months. The Foré were widely believed to be cannibals and their disease to be a consequence of their habit of eating dead relatives, as a mark of respect. It was the first spongiform encephalopathy to be shown to be infectious when Dr Carleton Gadjusek transmitted it by injecting material from cases into the brains of chimpanzees. His 1966 paper made an enormous impact. It was followed up two years later with the similar transmission of infectivity from sporadic CJD. Researchers have also been drawn to

CJD because the agent that transmits it from case to case has very unusual properties. To add to these distinctive qualities, it must also be said that the appallingly difficult scientific challenges it presents has also attracted researchers of character. The hardness of the questions posed by CJD have been matched by the strength of their personalities. This gave—and gives—the field the high profile that comes from the high levels of disagreement (and occasional vituperation) generated by invest-igators with strong views firmly held who are unafraid of challenging the consensus. Consider Hans Gerhardt Creutzfeldt. He described what he considered to be a new and unusual kind of brain disease in the early 1920s on the basis of clinical observations made when he was working in Alois Alzheimer's clinic in Breslau (then in Germany, but now, as Wroclaw, in Poland). Bertha E. was 22 when she came to the clinic. She had been brought up in an orphanage, and had developed problems with walking about six years before. Now she couldn't walk or stand without help. She had tremors and hallucinations, was incoherent with staccato speech, and did not know where she was. She rapidly became demented with muscle weakness and twitches, became mute, and started having epileptic fits which became continuous. She died about 12 months after first seeking medical help. Creutzfeldt described the autopsy findings and the microscopical appearance of her brain in his papers. For many years both Bertha's illness and the pathological find-ings were considered to be atypical of what came to be called Creutzfeld-Jakob Disease (CJD). However, all the features he described have been seen at one time or another in bona-fide CJD cases. But whether or not Bertha had CJD, it is right to continue commemorating his work by using his name. He was a remarkable person. He received his doctorate from the University of Kiel in 1909, became a ships doctor, and then trained in neuropathology and psychiatry. He was appointed professor at Kiel in 1925. Unlike many, if not most, of his medical colleagues, he was anti-Nazi. He was interrogated by the Gestapo on several occasions because he criticized Nazism in his lecture course, a very dangerous thing to do because university students tended to be keen supporters of Hitler. He allowed his clinic to be used as a refuge for people who had fallen foul of the 'hereditary laws', and his wife was imprisoned. After the Second World War he was responsible for discov-ering that a psychiatrist calling himself 'Dr Sawade' was in fact the noto-rious psychiatrist to the Gestapo, Dr Werner Heyde, who had been in charge of patient selection for the euthanasia programme from 1939 to 1941. An old Nazi who had taken part in the 'Kapp putsch' in March 1920 when the Erhardt Brigade free-corps with swastikas on their hel-mets had briefly occupied Berlin and caused the government to flee, Heyde owed his appointments to the influence of his friend

Theodor Eicke, murderer of Ernst Röhm, sometime commandant of
Dachau, and chief inspector of concentration camps. This is an inspectorate that will not be discussed further in this book. Heyde was arrested
after the war, but had escaped. He committed suicide in prison in 1961.

Any doubts about whether Bertha E. had Creutzfeld-Jakob disease
will have to linger forever because Creutzfeldt's microscope slides
have been lost. However, similar questions about Alfons Jakob's cases
were resolved when his slides were found in the archives of the
University of Hamburg's Neuropathology Department and re-examined
in the 1980s. One of the four cases that he described in 1921 fits the
modern definition. Ernst Kahn, a 42-year-old salesman and First World
War army officer, complained in May 1918 of aching legs, vertigo, and
abdominal discomfort. He then developed an unsteady gait, weakness in
the legs, double vision, and illegible handwriting. His mental condition
deteriorated, with loss of memory, disorientation, and hallucinations.
Periods of semistupor alternated with fearfulness and agitation. He had
painful sensations in the head and limbs. He became completely
demented and stuporose and died of pneumonia nine months after the
first symptoms had appeared. Like Creutzfeldt, Jakob had studied with
Alzheimer in Munich. He was a workaholic. After the First World War
he became one of Germany's leading neuropathologists as well as having a big private practice. He died in Hamburg in 1931 aged 47 after an
operation for an abscess in his abdomen that had developed as a complication of the chronic bone infection that he had had for many years.

Until the Second World War the Germans and the Austrians continued to lead the way in making discoveries about CJD. In 1928 Josef
Gerstmann described a new form of the disease. He worked at a mental hospital in Vienna and his patient was called Berta. She had been
completely healthy until, at the age of 25, she suddenly started staggering about. Her speech became slurred and to casual observers she
appeared to be drunk. Her moods swung from excessive cheerfulness
to irascibility and quarrelsomeness. She developed swallowing difficulties,
became demented, and died six years after the onset of her disease.
With Ernst Sträussler and Isaak Scheinker, Gerstmann reported her
case and its pathology in detail. Under the microscope the brain and
spinal cord had large numbers of what are now called 'amyloid
plaques'. These are microscopic spherical deposits of protein which
can be stained very strongly with a textile dye called Congo red so that
they stand out from other tissues. Berta's disease is now called
the Gerstmann–Sträussler–Scheinker syndrome. It runs in families;
17 affected members in five generations have been identified in hers.

Were all these discoveries made by German-speaking doctors
because CJD was commoner in German lands? Definitely not. It was

because more high-quality research was being done on brain diseases in those countries than in the rest of the world put together. Mad doctors there were not content just to run asylums or engage in private practice, as in Britain, but were active laboratory researchers and university professors as well. Bernhard von Gudden was a typical example of someone who did all these things. With one or two conspicuous exceptions, like Sir David Ferrier, who did brilliant experimental work on brain function in the 1870s in the first laboratory established in a British asylum (at Stanley Royd), people like Gudden had no British counterparts as clinician/scientists. There were hardly any professors of psychiatry in England until after the Second World War. Another advantage of the German system was the very close nature of the links between psychiatry, neurology, and pathology. In Britain relationships between them were never strong and for the last hundred years and more the subjects have developed as independent specialities with strong boundaries between them. Neurologists regard themselves as intellectual physicians. Even if they cannot treat the diseases they diagnose, which is usually the case, they can pinpoint the seat of the mischief with anatomical accuracy. They look down on psychiatrists because they cannot do this. Both regard themselves as superior to pathologists because they are clinicians dealing with live patients. This may be a caricature. But there is truth in it. There can be little doubt that the existence of separate specialties concerned with brain diseases has not aided the diagnosis and management of variant CJD.

Amyloid plaques in the brain are a distinctive feature of the Gerstmann–Sträussler–Scheinker syndrome. They are also abundant in patients with variant CJD. Together with spongiform degeneration—the appearance of holes (vacuoles) in neuronal processes (the spiky outgrowths that join nerve cells together)—nerve cell death, and a reactive astrocytic gliosis (an increase in the number and size of the cells in the brain that surround nerve cells which occurs in response to nerve cell death) they define the pathological diagnosis of CJD. Amyloids are accumulations of protein fragments characterized by the way they react to special stains. They build up in the brains of people with Alzheimer's disease. Before the antibiotic era patients with chronic infections such as tuberculosis and osteomyelitis often died of renal failure because amyloid built up in their kidneys. The amyloid was a by-product of the massive amounts of immune defence proteins they had produced to fight the causative microbes. Although they stained in the same way, the proteins in these cases were different from the one found in CJD, however. In CJD it is built up of chunks of an abnormal protein called PrP^{sc}.

It is the build up of PrP^{sc} in the brain that causes CJD. This is particularly intense in variant CJD, where it accumulates in two kinds of plaque. 'Florid' ones are collections of amyloid located at the centre of vacuoles. 'Immature' plaques lack the special staining properties of amyloid. Both kinds of plaque are very common in variant CJD. PrP^{sc} is also found in the tonsils, lymph nodes, spleen, and appendix.

PrP^{sc} is an abnormal form of a protein called PrP. We all make PrP continuously in many different cells in our bodies, particularly in the brain. The individual molecules do not survive very long. For obvious technical reasons, their fate in normal human brains has not been worked out, but laboratory studies on tissue culture cells shows that once made, they only last a few hours before they are digested by enzymes called proteases and scavenged away. Their rate of production is the same as the rate of destruction, so despite a rapid turnover of the molecules, PrP levels are evenly maintained. But PrP^{sc} is resistant to protease digestion. This is why it builds up in CJD.

Proteins are long folded-up chains made up of different amino acids linked together. Their basic properties are determined by the linear sequence in which the amino acids are arranged. The PrP^{sc} in variant CJD has exactly the same sequence as normal human PrP, but the shape of the two molecules is not the same. Their chains are folded in very different ways. This is why PrP^{sc} is resistant to digestion. Understanding how PrP^{sc} is arranged in this way is the central question in CJD research because there is a strong scientific consensus that PrP^{sc} is the main villain of the piece. Without its formation there would be no disease.

Chapter 9

The Science of TSEs

It is gratifying that such a complex disease as CJD has such a seemingly simple explanation. But to reach it scientists have trodden several very different and complicated pathways. Their journeys have been difficult and slow. They are far from finished. Much of the knowledge about transmissible spongiform encephalopathies (TSEs), the family name for diseases such as CJD, comes from work on scrapie. Scrapie is a disease of middle-aged sheep. It has been widespread in Europe for centuries ever since animal diseases were first systematically recorded. It was described in British sheep in 1732, in Germany (as *traberkrankheit*) in 1750 and in France (as *la tremblante*) in 1810. In Hungary it is called *súrlókázjánka* and in Iceland *rida*. Its onset is insidious and its clinical progress relatively slow. In the beginning there is mild impairment of social behaviour. Animals become restless, apprehensive, hyperexcitable, and aggressive. They develop fine tremors of the head and neck, which go on to be more generalized, affecting the whole body and producing a shivering effect. Starting at the root of the tail, the animal experiences intense sensations in its skin which cause it to rub and scratch with great vigour. Thought to be due to itching, this is usually a very characteristic feature of the disease, although it was not seen much in rida. Walking becomes affected with an unsteady and uncoordinated gait, the animals moving with high-stepping or stiff front legs, holding a braced stance, or swaying and leaning to one side. Eventually they become emaciated, stuporous, and blind, and they fall down. Not all animals show all these manifestations. Like any concise description, this one is a compromise. It only highlights common features. Sometimes sheep just drop down dead and can only be diagnosed at post-mortem. In others the gait problems predominate, and in many it is the itching and scratching. Nevertheless, all of them die. In the 1930s French researchers showed that scrapie could be transmitted from sheep to sheep by the injection of brain material. The incubation period was more than a year, roughly a hundred times longer than for most bacteria and viruses. An even more unusual feature of scrapie was discovered by accident at about the same time. Loup is a Scots word meaning to leap or spring, and louping-ill is a brain disease of sheep caused by a virus transmitted by ticks. It occurs in Northern England, Scotland, and Ireland. The animals develop a peculiar jumping gait. In the 1930s a veterinarian working at the Moredun Research Institute in Edinburgh, W. S. Gordon, developed a vaccine against the virus. He infected animals, waited a few days for the virus to grow, and then collected brain, spinal cord, and spleen and killed the virus by treatment with formaldehyde. Formaldehyde is a pungent chemical which reacts with proteins and nucleic acids, particularly with the exposed parts of the molecules. If allowed to react with them long enough it causes permanent changes by forming strong chemical bonds between the parts of the molecules that are close together. The trick in vaccine manufacture is to find the concentration of formaldehyde and reaction time that will kill the microbe, or inactivate the protein toxin, but leave them sufficiently unchanged chemically so that they can still stimulate the immune system to make a response that will protect against infection. Gordon treated his louping-ill infected tissues with 0.35 per cent formalin for at least three months. These harsh conditions killed the virus with a big margin to spare. However, one batch of vaccine was made from sheep that were incubating scrapie. Within two to three years more than one-third of the 18,000 animals given this batch had developed scrapie, a very much higher incidence than expected in these particular animals. Gordon published the details of this event in 1946; he confirmed the transmissibility of scrapie from sheep to sheep by injection of brain material in other large experiments as well. Gordon's work was important for several reasons. It was the first demonstration that there was something really unusual about the scrapie agent, because the strength of resistance to formaldehyde that it showed was quite unprecedented. It also confirmed the long incubation period of the disease. This was unusual, but unlike complete resistance to formaldehyde, not unprecedented. Diseases like rabies can also have very long incubation periods. Gordon's cohort of immunized sheep was a mixed bag of breeds, and the third distinctive thing was that some seemed to be more resistant to infection than others. The reasonable assumption was that these differences had a genetic basis. Another rather unusual property of scrapie as an infectious disease that the louping-ill vaccine event highlighted was that even if scrapie was naturally present in a flock, in contrast to the 30 per cent and more that came down after the immunization, the disease usually affected only a small proportion of animals, often 1 per cent or thereabouts. In practice this diminishes the incentive for shepherds to take aggressive control measures. The economic loss from the occasional animal dying now and then is not particularly great. Since the Second World War the disease has had its biggest direct financial impact in countries that have adopted an eradication programme requiring the slaughter of affected flocks. The USA spent $3,000,000 on indemnity payments to flock owners between 1947 and 1964 in an eradication programme after the introduction of the disease into Michigan in 1947, supposedly in imported British sheep.

The most significant aspect of the louping-ill vaccine event was that it was the first time that a TSE had shown its iatrogenic teeth. It demonstrated two very important things, that by-passing body defences by injection could give the transmission of an infection an enormous degree of help, and that there were agents of disease in nature, particularly in animal brains, that did not follow the rules. It is easy to point these things out with the benefit of hindsight, and it has to be said that Gordon's louping-ill vaccine work did not have a big impact when it was published. Even now, it receives less attention than it should. Medicine of course has been compelled to learn the first of these lessons over and over again in the last half-century and more. Hepatitis viruses and blood transfusion, and intravenous drug abuse and HIV are classic examples. Transmission of these viruses through the injection 'by-pass' has, of course, made tough lives even tougher for haemophiliacs and heroin addicts.

A big methodological breakthrough in scrapie research occurred in 1961, when Dick Chandler of the Institute of Animal Health at Compton, Berkshire, transmitted scrapie to mice. The incubation period of infection was reasonably short, often months, rather than the years in sheep. Now researchers had a reasonable prospect of using standard laboratory approaches to find out what the scrapie agent was and how it worked.

The year 1961 was a good time to make this technical advance, because it happened when animal virology was going through a period of revolutionary change driven by discoveries made by molecular biologists and by the introduction of enormously powerful laboratory methods.

A particularly important development because of its impact on thinking about how infectious agents were constructed was the simultaneous demonstration by rival groups in Berkeley, California, and in Tübingen, Germany, in 1956, that the infectivity of tobacco mosaic virus (TMV) resided in RNA and not in protein. This rod-shaped virus had been under experimental scrutiny for a long time. It was an excellent model to study because it could be grown in tobacco plant leaves in very large amounts. It was economically important and so funding for its study was not difficult to get. The TMV RNA discovery added strength to the conclusions of the Hershey–Chase experiment. For a long time it had been thought that the protein which made up most of the structure of the virus particle was infections. Francis Crick took the demolition of this notion as illustrating his 'Central Dogma' very well, that 'once information (meaning here the determination of a sequence of units) has been passed into a protein it cannot get out again, either to form a copy of the molecule, or to affect the blueprint of a nucleic acid'. The demonstration of the infectivity of its RNA was not the first time that a historically important discovery had been made independently by two different TMV researchers. At the end of the nineteenth century Dimitri Ivanovskii in St Petersburg and Martinus Beijerinck in Delft showed that its infectivity passed through filters that would hold back bacteria. This was the first time that a physical method had been used to operationally define an infectious agent. It was used later to do the same for scrapie. Beijerinck and Ivanovskii fell out about who was first. Sixty years later relations between the discoverers of RNA infectivity, Heinz Fraenkel-Conrat in Berkeley and Gerhard Schramm in Tübingen, were not good either. But the main thrust of their argument was about the detailed chemical structure of the protein coat of the virus and the methods used to determine it. Fraenkel-Conrat thought Schramm was a sloppy chemist. Schramm did not care. Under the circumstances it is a curious fact that their antipathy was only a professional one. Both were Germans. Fraenkel-Conrat was Jewish, the son of a well-known obstetrician in Breslau. He had qualified in medicine but left Germany and obtained a Ph.D. in biochemistry from Edinburgh in 1936 before going to Brazil and the USA. Schramm never left Germany. Not only did he join the Nazi party, he was a member of the SS.

The work on TMV would not have been possible without new developments in laboratory equipment and techniques. They also had a very big effect on animal virology. The ultracentrifuge was particularly important. By spinning plastic test tubes in tough metal rotors in an armour-plated cage at speeds of 40,000 revolutions per minute, enormous gravitational forces could be developed, forces great enough to cause very small virus particles to sink down a column of liquid either to form a pellet at the bottom, or if the fluid in which it was suspended had the right specific gravity, to come to rest at a particular position in the tube separated from impurities with different densities. Ultracentrifuges were developed before the Second World War but did not become commercially available at reasonable prices until the 1950s. Their impact was nicely described in 1967 by the virus biochemist Dean Fraser:

> In my opinion this machine, the Spinco ultracentrifuge...has done more than any other single instrument to advance the study of viruses.... Previously, ultracentrifuges had to be run in the subbasement behind three-foot-thick walls of reinforced concrete. They were individually built; the scientists who operated them were considered slightly insane. Centrifuges required constant attention... a willingness to spend most of your time flat on your back in a puddle of oil doing plumbing and mechanical repairs. Explosions that wrecked the machine were to be expected at fairly regular intervals. The Spinco preparative centrifuge, for contrast, is about the size of a family washing machine, and anyone can learn to run it in ten minutes.

Using the ultracentrifuge it became possible to concentrate and purify virus particles for biochemical analysis. It also made seeing them in the electron microscope a seasonably straightforward task. For this work the most important technical development, after the invention and development of the microscope itself, was the introduction of negative staining, a remarkably simple and rapid method in which viruses, which cannot be seen on their own because electron pass right through them are visualized as negative images by being embedded in a solution of electron-opaque heavy metal salts. This technique was introduced into routine use in 1959 and led almost overnight to a massive increase in knowledge about the precise details of virus structure.

All this work was very relevant to scrapie. It was not caused by a bacterium because it was too small to be one. Filtration experiments like those done by Beijerinck and Ivanovskii had conclusively shown that. But ultracentrifugation of brain extracts did not give the results expected for a virus. In centrifuge tubes containing liquid at different densities, infectivity was left diffusely spread through the column of liquid rather than being concentrated at a particular place. If it was associated with regularly shaped small particles in suspension, either they did not behave according to physical rules, or infectivity resided in something else. However, with the right centrifugation conditions it was possible to concentrate infectivity at the bottom of the centrifuge tube. But negative staining and electron microscopy of the pellets only revealed fragments of brain cells. No virus-like particles were seen. So other, more indirect, methods had to be used. A range of physical and chemical treatments was applied. The conclusion had to be drawn from their results that the agent of scrapie was not a conventional virus. Boiling kills the latter very quickly, but scrapie very slowly. Infectivity can survive the conditions of heat and pressure that are used in hospital autoclaves set to kill the most heat resistant agents normally encountered in medical practice, bacterial spores like the ones responsible for botulism, tetanus, and gas gangrene. Tests using chemicals did not give a clear indication about what the agent was made of. Its unusual nature was also confirmed by its extreme resistance to radiation, whether gamma rays from cobalt 60 or ultraviolet. This meant that if it contained a nucleic acid, it was a very small one. A typical and important conclusion was drawn by the South African radiobiologist Tikvah Alper and her colleagues in London in 1967. They blasted infected mouse brains with a very powerful ultraviolet ray generator: 'the results confirm our previous conclusion that scrapie is most unlikely to depend on a nucleic acid moiety for its replicative ability. There is, however, no conclusive evidence from these experiments as to whether or not the agent might be associated with a protein...' These conclusions were reached using calculations based on target theory. Although first developed in the 1920s this approach to characterizing genes had been put on the map in a big way in a paper written by Max Delbrück years before he became leader of the 'phage church'. In 1935, he was still a physicist working at the Kaiser Wilhelm Institute for Chemistry in Berlin. He had become interested in biology, however, and with the Russian geneticist Nicolai Timoféeff-Ressovsky, who ran the genetics laboratory at the Kaiser Wilhelm Institute for Brain Research, and an expert in measuring radiation doses, K. G. Zimmer, published a quantum mechanical model of the gene. Although it turned out to be wrong, it was enormously influential in stimulating all sorts of scientists to become molecular biologists. Published in a journal that was hardly read by anyone, the 'Nachrichten den gelehrten Gessellschaften der Wissenschaften zu Göttingen', it became famous by accident. Copies of the paper were sent to those the authors hoped to impress. The reprints had a green cover. One was sent to a crystallographer in Belfast, P. P. Ewald. He passed it on to the famous Austrian physicist Erwin Schrödinger, who in 1938 had been recruited by Eamonn de Valera to work in Dublin on theoretical physics. In 1944 Schrödinger wrote a little book called *What is Life*. In it he discussed Delbrück's model of the gene. He said 'from Delbrück's general picture of the hereditary substance it emerges that living matter, while not eluding the "Laws of Physics" as established up to date, is likely to involve "other laws of physics" hitherto unknown . . .'. It is not surprising that the book stimulated many scientists, particularly physicists returning from war work, to turn to biology as the coming thing, and it gave the 'Green Pamphlet', as Delbrück's paper was known and its author a big reputation. (The opposite happened to Timoféev, who had stayed in Berlin during the war. He was captured and condemned as a traitor by the Soviets, but survived because his radiation expertise was useful to Beria's atom bomb project.)

When Delbrück was still a physicist in Berlin, he worked for Lise Meitner, the first woman to be appointed to a professorship in physics in Germany. She had received the title 'ausserordentlicher Professor' at the Kaiser Wilhelm Institute of Chemistry in 1926. In a curious coincidence, Tikvah Alper had also worked for Meitner. She had been one of her students between 1929 and 1932. It is unlikely that Delbrück's and Alper's paths crossed in Berlin because Delbrück did not start work as Meitner's theoretical assistant until the autumn of 1932, and there is no evidence that Delbrück's thinking about target theory directly influenced Alper's work. But both had been lucky to work for Meitner. Association with an outstanding scientist is the best way to become one, and they had chosen wisely. Meitner's physics was of Nobel quality, but she never received the prize. A recent study of Nobel committee records by Friedman has shown how her fate was in his words 'sealed by Swedish scientific leadership's insensitivity and self-interest'. It was an example of how the search for truth rarely transcends political realities and personal prejudices (Figure 9.1).

Alper's work on the scrapie agent provided good evidence about what it was not, and so raised very interesting questions about what it was. However, it did not, and could not, answer them. Progress could only be made by studying it in a different way. It is a sound rule in biology that if one has problems studying structure, then you turn to function. A good way to do this is to use genetics. This was how Delbrück and his 'phage church' constructed one of the main foundation stones of molecular biology. Genetics done his way turned out to be an incredibly powerful way to answer big biological questions, even if there was personal disappointment for him because no new laws of physics emerged. Speed of experimentation was one of the reasons why phage was chosen. A genetic cross could be completed and its results analysed within 24 h. But even with Chandler's mice, genetic experiments with scrapie could regularly take more than 600 days.

Figure 9.1 Solvay Congress on Atomic Physics, Brussels 1933. Sitting: 1st from left, E. Schrödinger (p. 129); 3rd from left, N. Bohr (pp. 112, 176); 2nd from right, L. Meitner (pp. 130, 132, 176, 180, 200). Standing: 4th from right, R. Peierls (p. 176–177); 5th from right, J. D. Cockcroft (p. 176).

In some experiments there was a reasonable suspicion that mice had died naturally of old age while still incubating the disease. As an impediment to progress, this set scrapie geneticists a very serious challenge. Another very big obstacle for them was the extremely small number of variable properties possessed by scrapie agents. In comparison with Gregor Mendel's peas, with all the different appearances that could easily be scored by eye, like colour, wrinkledness or smoothness, and long or short stems, there were hardly any. All they had to start with was the length of the incubation period.

Unlike Delbrück's bacterial viruses, scrapie could not be grown in free living cells in the laboratory. It produced no visible effect on sheep or mouse tissue culture cells—the equivalent of *E. coli* for bacteriophage. Even when animals were known to be infected they did not respond by producing antibodies. None could be detected by any of the sensitive tests that were available in the 1960s. This apparent complacency of the host to infection also applied to its inflammatory response. There was not one. Bacterial meningitis or viral encephalitis can be diagnosed by an increase in the number of white cells in the cerebrospinal fluid, the liquid surrounding the brain and spinal cord. This never happened in scrapie despite the extensive destructive changes that went on in the brain. All these things meant that the scrapie geneticists could not speed up their experiments by making early diagnoses using blood tests. They had to wait for the animals to die. Unfortunately, the same applies to all TSE diseases, not just experimental infections. Its practical impact for making a diagnosis in life is the same. It is very difficult.

Money and the right people were needed for the genetic attack on scrapie to be successful. Long-term funding was essential. In the 1950s, when the work started, there was still a belief that curiosity-led research initiated on the basis of the personal predelictions of individual scientists was a powerful way of answering scientific questions and making discoveries of practical importance. Maybe this was because examples like the discovery of penicillin by Alexander Fleming and nuclear fission (made by Delbrück's physicist mentors in Berlin, Otto Hahn and Lise Meitner, just before the Second World War) were still fresh in the minds of policy makers. This philosophy meant that funding was available from the Treasury. Short-term successes would be welcome, but were not necessary; money was provided as a long-term investment. Performance indicators were a thing of the future.

High-quality scientists with patience were also needed. In 1954 Alan Dickinson got his Ph.D. in Genetics from Birmingham University, and joined the staff of the Animal Breeding Research Organisation (ABRO) in Edinburgh. He was given the personal freedom to follow his own research lines. At about the same time J. T. Stamp was appointed Director of the Moredun Research Institute, also in Edinburgh. The institute specialized in sheep diseases, and ever since Gordon's work on the louping-ill vaccine, had been interested in scrapie. Stamp's predecessor had even tried to make a vaccine; the unsuccessful attempts confirmed its extreme resistance to chemical and physical inactivation. Stamp contacted ABRO because he wanted to work on scrapie genetics, and set up a collaboration with Dickinson. For the first few years all the work was done with sheep. Although very slow, it was necessary. After all, scrapie was a sheep disease, and it was already known that different breeds seemed to show different susceptibilities to it. There was even a feeling among some sheep breeders that it could be inherited.

Dickinson spent the rest of his career working on scrapie (Figure 9.2). His research papers are models of clarity. He showed that although the relationship between scrapie and its mammalian host was a surprisingly simple one, it was unlike that found in any other animal infection. His work, with that of his closest Edinburgh colleagues, laid down the basic ground rules for the biology of TSEs. Without it we would still be arguing about the relationship between BSE and variant CJD. But Dickinson's professional interests were not restricted to the arcane genetical world of murine and ovine alleles, genotypes, and

Figure 9.2   Alan Dickinson, 1983.

overdominance. In October 1976 he warned that using human pituitary glands collected at post-mortem to make growth hormone to treat children with growth problems carried a very real risk of transmitting CJD. He made practical recommendations about reducing this risk. Sadly, not all of them were followed. The first British victim of these circumstances died on 16 February 1985. She had been infected before Dickinson's warning. But cases followed which could have been prevented.

Dickinson's achievements appear to be pretty straightforward. But it would be wrong to describe either his thought processes or his research in this way. They are best characterized, in the best possible sense, as intense and convoluted. But only someone with these attributes could have grappled successfully over the years with the complexities of scrapie in mice and convincingly reduced its genetics to straightforward essentials. Dickinson illustrated it himself when giving oral evidence to the BSE Inquiry about a very important paper that he published in 1971 which drew simple conclusions from an incredibly complicated matrix of results:

> What I had done is had got to the point of doing all the massively complex calculations, which had to be done by hand, a year earlier. I said 'This is so complex I will put it away for a year.' I put it away for twelve months then got it out, started out, having forgotten all the detail,

worked it out and got exactly the same answer. We published it . . . What happened next was slightly shocking. About a year later the Professor of Genetics then at the Institute of Genetics in Edinburgh, Douglas Faulkner, came to me and said: 'You are going to give a talk on scrapie next week, would you mind not giving it?' I said. . . . 'OK'. He said 'I will have to let you into a confidence. We have set the paper you published as the exclusive contents of the three hour final exam in genetics. . . .' I said 'It is far too hard. . . . It is far too difficult.' And of the 17 people who took it, 14 recorded, at great length, that it was infinitely too difficult. It really flummoxed them.

In August 1973 Dickinson attended the 13th International Congress of Genetics at Berkeley, California. He gave a paper summarizing his discoveries. He explained how it was possible to do scrapie genetics in mice without being able to see the agent that caused the disease or even without having any clear idea as to what it was—in biochemical jargon, its 'macromolecular composition'. He said that to do this kind of work it had been necessary to control eight factors. Four were in mice: their genetic make-up, their age, their sex, and how they were infected; and four were to do with the scrapie agent: its strain, the dose given to the mice, the genetic make-up of the mice in which it had been grown, and the mouse organ from which it had come. The two properties that he studied were the incubation period of the disease, and the kind and distribution of damage produced in the brain (the 'lesion profile'). Before describing his genetic findings he pointed out that when the eight factors were held constant the predictability of these was remarkable. It was 'greater than has yet been found in any other class of disease'. He went on to remind his audience that biochemical attempts to purify the agents had not been successful so

> it is fortunate that a geneticist does not need to know what types of macromolecules carry the information if he has different strains of agent which can be measured and handled experimentally on condition that these strain differences stem from the agent and not merely from the host. We do have such agents.

He then described his main genetic discovery, which was that a mouse gene controlled the incubation period of the disease. He called it *sinc* (an acronym for *s*crapie *inc*ubation period). There were three kinds (genotypes) of *sinc*, s7s7, p7p7, and s7p7. (As in people, most mice genes occur in pairs which can be identical (homozygotes), or not (heterozygotes).) To find a gene that controlled the outcome of an infection was not particularly surprising. People had been speculating about it for many years. It had long been held, for example, that Scots with red hair were more susceptible to tuberculosis. But to find definitive genetical proof of such a relationship, rather than just a mythical story, was a breakthrough, particularly in animals. The only apparent precedent was in plants. In the 1940s the plant pathologist Harold Flor had shown using flax and flax rust, a fungus that attacked it, that there was a tit-for-tat (gene-for-gene) relationship between a disease resistance gene in the flax and a virulence gene in the fungus. Disease could happen if the first was not working well, or, if it was, when the second mutated to overcome it. Since then many other plants and plant pathogens have been found to show the same relationship. But it is quite different from the one between *sinc* and scrapie. Different genotypes of *sinc* give completely different results with different scrapie strains. Thus s7s7 with some strains gives a much shorter incubation period than p7p7, whereas with other scrapie strains the result is the exact opposite. Likewise, the fate of s7p7 (heterozygote) mice also depends on the type of scrapie strain. With some of the incubation period is longer than for either the p7p7 or the s7s7 homozygote. I am sure that many readers of this book will find these observations hard to interpret. They certainly 'flummoxed' most of the honours class in genetics at Edinburgh University. But they enabled Dickinson to make predictions which have stood the test of time. In particular, he said that they indicated that the replication of the scrapie agents depended on the *sinc* gene product. Implicit in this statement was another prediction, that if the *sinc* gene is removed, there will be no product, and so such an animal would be completely resistant to infection with scrapie. Twenty years elapsed before this was put to the test, when transgenic mice (knockouts) were made in Switzerland with a defective *sinc* gene. By this time it was known that the *sinc* product was the same as PrP. These mice did not make it, and they were resistant to infection with scrapie, confirming the central role that PrP plays in the disease. This is one of the clearest results that has been obtained in TSE research. But it held one big surprise. The PrP protein or its gene has been found in all vertebrate species tested, including birds. It is made in quite large amounts in the brain, heart, and muscles. It would be reasonable to guess that animals without it would be disabled in some way. But two of the four independently and differently constructed kinds of knockout mice that lack it have normal lives. These are the Swiss ($Prnp^{o/o}$[Zurich]) and Scottish ($Prnp^{-/-}$[Edinburgh]) lines. On the other hand, Japanese knockout mice ($Prnp^{-/-}$ [Nagasaki]) start staggering about when they are about 70 weeks old due to the loss of nerve cells in their cerebellums; because of their genetic make-up it is possible that their problem has nothing to do with PrP. So the big puzzle is what does PrP do? The question remains unanswered.

The Dickinson mouse genetic approach to scrapie has been crucial to TSE work, not only because it laid down important principles and generated testable hypotheses, but because it generated a highly reproducible and very robust typing system. The importance of this cannot be overemphasized. Without it researchers on infections caused by TSEs would be like detectives trying to identify criminals without fingerprinting or DNA. Despite being developed as long ago as the early 1970s, and despite having not even a whiff of molecular biology about it, the combination of incubation period and lesion profile worked out by Dickinson and his colleagues Hugh Fraser and Moira Bruce remains the gold standard method for identifying strains. That the agents from variant CJD and BSE cases are identical in these tests is the main piece of direct evidence supporting the conclusion that both are caused by exactly the same thing. Everything else—the timing of the cases and their commoness in Britain and rarity elsewhere—is circumstantial.

By 1972 Dickinson's research programme in Edinburgh had delivered important insights into scrapie biology. But TSE research was about to undergo a change of emphasis. In September of that year a patient with CJD was admitted to hospital in San Francisco. The neurology resident was Stanley Prusiner. Although medically qualified, he had already done a good deal of basic laboratory research. He had spent three years studying enzymes in *E. coli* at the National Institutes of Health. This had made him into a card-carrying biochemist. So his approach to biological problems was very different from Dickinson's. For Prusiner the challenge was to find out what the scrapie agent was made of. His tools were the test tube, the electrophoresis kit and the ultracentrifuge, rather than the calculating machine. He set up his scrapie laboratory at the University of California in San Francisco in July 1974. His early work focused on whether the infectious part of the scrapie agent was made of protein or nucleic acid, and by 1982 he had concluded that the latter were not involved. By brute force biochemistry he succeeded in purifying the infectious agent 3000 to 10,000 fold. This made it possible to establish the amino acid sequence of the single protein that remained. It was PrP^{sc}. Antibodies were made against it, and they reacted with the amyloid plaques in CJD, confirming a central role for this protein in the disease. Comparison of PrP^{sc} and PrP showed that they had exactly the same sequence, despite their very different behaviour in the test tube. PrP^{sc} molecules were very resistant to digestion with protease enzymes whereas PrP was very sensitive to them. Knowing the sequence of the PrP gene, which every human has, made it possible to look at the PrP genes in families that suffered from inherited forms of CJD. This brought further confirmation of its importance in the disease. In 1989 a family with the Gerstmann–Sträussler–Scheinker syndrome was found to have a particular mutation in their PrP gene (a CCG to CTG mutation at codon 102 causing a leucine for proline substitution). Many other examples of PrP mutations linked to CJD have been found since then.

This biochemical work by Prusiner was of high quality. Like Dickinson's genetics it had benefited enormously from the intensity of his approach. It was painstaking and very thorough. It used all the best state-of-the-art techniques, and he collaborated with other scientific groups of top standing, like the one in Zurich led by Charlie Weissman. None of its findings have been seriously disputed. It still goes on. It has led to a consensus that conversion of PrP to PrP^{sc} is the central event in the causation of CJD and the other TSEs. Nobody now doubts that the agent responsible for this event and for infectivity is different from conventional microbes, like viruses, and so a majority of scientists subscribe to Prusiner's 'prion' hypothesis.

Prusiner introduced the term 'prion' in a long paper that appeared in one of the world's premier scientific journals, *Science*, in 1982. He said

> In place of such terms as 'unconventional virus' or 'unusual slow virus-like agent', the term 'prion' (pronounced *pree-on*) is suggested. Prions are small *pro*teinaceous *in*fectious particles which are resistant to inactivation by most procedures that modify nucleic acids. The term 'prion' underscores the requirement of a protein for infection; current knowledge does not allow exclusion of a small nucleic acid within the interior of the particle.

Since then its definition has changed. It now reads 'prions are defined as proteinaceous infectious particles that are devoid of nucleic acid and seem to be composed exclusively of a modified isoform of the prion protein designated PrP^{sc}'. Despite the illogical circularity that defines an object as a modified form of itself, in Prusiner's own words 'a semantic conundrum', in operational terms everyone knows what it means. It is, for example, the agent produced by cows with BSE that has gone on to infect people and cause vCJD.

There is a minority of scientists, however, who do not go along with the more recent definition. They agree wholeheartedly with one of the main conclusions that Prusiner drew at the end of his 1982 paper:

> The importance of prion research in the potential elucidation of a wide variety of medical illnesses underscores the need for purification of the scrapie agent to homogeneity and the subsequent identification of its macromolecular components. Only then can we determine with certainty whether or not prions are devoid of nucleic acids.

One essential problem for them is that although Prusiner has done brilliant work in purifying PrP^{sc}, it is so inefficiently infectious that it is very difficult to rule out that some other component is necessary as well. His own figures show that although one milligram of purified prion contains 60 billion infectious units, if one of these units has a molecular weight of 30,000—the same as PrP^{sc}—then infectious units only make up 0.0003 per cent of the purified preparation. Proving conclusively that these needles in the haystack are biochemically identical to the other 99.9997 per cent PrP^{sc} molecules—and thus ruling out the presence of, say, an associated small nucleic acid—is close to impossible.

It is easy to criticize Prusiner's 1982 'prion' paper. It is selective in its use of evidence to support the idea that infectivity resides in protein and not nucleic acid. The first unusual property of the scrapie agent that it mentions, in its second, introductory paragraph, is the remarkable resistance to formaldehyde revealed by the Gordon louping-ill vaccine episode. But although this chemical has long been known to react with proteins and inactivate their function (it was being used to inactivate diphtheria toxin to make a vaccine in the 1920s) it is not mentioned again. It is a tribute to the influence of the 1982 paper that a lack of interest in seeking an explanation persists to the present day, and the resistance of TSEs to it remains a mystery. One possible explanation could be that formaldehyde does not react with prions because they lack the right chemical structures. This can be ruled out, not only because the sequence of the protein indicates that there should be plenty, but because treatment with it *protects* infectivity almost completely from heat inactivation. This bizarre and unexpected result not only shows that a reaction has taken place, but like formaldehyde resistance itself, does not fit easily with the prion (protein only) hypothesis.

Much of the argument in the paper about the possible role of nucleic acids focuses on the effect of treating the scrapie agent with radiation and chemicals. Its resistance to the simple chemical hydroxylamine is given as one piece of evidence supporting the absence of a role for nucleic acids. It does nothing of the kind. This is the one aspect of the 1982 paper on which I have direct experimental experience. In the 1960s I worked on the effect of hydroxylamine on conventional viruses. All had nucleic acid genomes essential for their infectivity and some were sensitive to it—they were easily killed. But some were completely resistant, as the Tübingen virologists had already shown tobacco mosaic virus to be in the early 1960s. However, neither hydroxylamine nor formaldehyde, irritant as these chemicals may be, were the reasons why the 1982 paper created such a stir. Prusiner himself says in it that one of its main proposals was 'clearly heretical'. The proposal was that the scrapie agent is an infectious protein. In fact, although he put forward two proposals to explain how a protein could be infectious, only one of them fell into this category. The uncontroversial idea was that the scrapie prion acted as a switch to turn on an inactive prion gene in the host. When activated it made more prions. This proposal predicted that there would be no prion product in the cells of normal hosts. The heretical idea was that the infectious prion coded for its own synthesis. It was heretical because it contradicted the 'central dogma' of molecular biology. But Francis Crick's statement that once sequence information had been passed into a protein it cannot get out again to form a copy of the molecule was not an *ex-cathedra* decree based on faith. After all, he was an atheist as well as a scientist. Rather, it had come from 40 years of intense study. It also rested on the firmly established chemical rules governing molecular interactions. So, if Prusiner was right, it would be back to the drawing board for both biologists and chemists. Delbrück had given up phage work in 1953 just before Watson and Crick published their double helix paper because molecular biology was in good hands and he thought that its basic principles were becoming so firmly established that soon nothing would be left but filling-in details. But had he been wrong? Was his long dead hope that studying biology might lead to the discovery of new laws of physics about to be resurrected? Within three years Prusiner's own biochemical work had shown that Delbrück's ghost could rest in peace. Both of the prion (protein only) proposals in the 1982 paper turned out to be wrong. The scrapie prion neither switched on a cellular prion gene or coded for itself as an infectious agent. It was instead the abnormally folded form of a protein continuously made in normal cells—so there was no need for a switch mechanism or unconventional coding process. The 'central dogma' was safe. Dickinson's 1973 suggestion that the 'scrapie agent is a deviant form of some normal cell constituent and that its replication depends on the same subcellular systems that support production of the normal homolog' had turned out to be right.

Prusiner got into a lot of trouble because of his heretical hypothesis. But his 1982 paper turned out to be a rhetorical success. The term 'prion' has become as firmly established as 'gene', or 'species'. But not everyone is sure exactly what it is. My own view is that there is enough uncertainty to justify keeping an open mind. One big obstacle to universal agreement that PrP^{sc} is the sole component of the infectious particle is the existence of different strains. The protein-only prion hypothesis currently in vogue says that strain type is determined by the way PrP^{sc} is folded. This implied that for a TSE like scrapie, in which Dickinson and his Edinburgh successors have identified different types numbering well into double figures, there must be the same number of folding patterns. Although protein folding has been studied intensively for more than half a century, nobody knows whether this is possible, because the rules that govern protein folding are still incomplete. Currently they only give general guidance. One comfort factor is that the PrP to PrP^{sc} conversion—about which there is no controversy—has itself caused a new rule to be written. Until this conversion was discovered, it was universally agreed that a protein would only adopt one structure, with at most minor flexibilities, and that the pattern of folding would be entirely determined by its amino acid sequence. But the change from PrP to PrP^{sc}—which, of course, have identical sequences—is a dramatic and fundamental one. Major features of the folding of PrP are alpha helices in which the protein chain is arranged in long coils. In PrP^{sc} the dominant structure are β-sheets, which hardly occur at all in PrP. As their name indicates, they are not coiled. The first clue that PrP^{sc} had β-sheets in its structure came from the amyloid deposits found in the brains of CJD victims revealed by special stains like Congo red. There are many different proteins in the amyloid family, but the common feature in all of them, and the one responsible for their characteristic interaction with Congo red, is the presence of β-sheets. A major obstacle to working out the precise details of PrP^{sc} structure and finding out whether strain differences are due to variation in it is its stickiness. Until the technical difficulties of working with it are overcome, however, sceptics will continue to doubt whether PrP^{sc} can adopt all the structural forms demanded by the existence of multiple strains. Dickinson has proposed that a small nucleic acid associated with PrP^{sc} plays a role in all this. He has jokingly called it the 'virino'. But like its analogue, the weightless 'neutrino', it has yet to be isolated in a test tube. My own view is that tiny RNAs may turn out to be involved. Cells have lots of them but they are difficult to study and so have been described as the biological equivalent of dark matter—all around us and important but difficult to detect.

The protein-only prion hypothesis says, in Prusiner's words, that the 'mechanism by which PrP^{sc} is formed must involve a templating process whereby existing PrP^{sc} directs the refolding of PrP into a nascent PrP^{sc} with the same conformation'. The hypothesis is silent about how this happens. It is its major weakness. It harks back to the 1930s and to theories about protein synthesis that were put to rest by a short paper published by Linus Pauling (the discoverer of the alpha helix) and Max Delbrück in 1940. They said:

> It is our opinion that the process of synthesis and folding of highly complex molecules in the living cell involve, in addition to covalent-bond formation, only the intermolecular interactions of van der Waals attraction and repulsion, electrostatic interactions, hydrogen-bond formation, etc, which are now rather well understood. There interactions are

such as to give stability to a system of two molecules with *complement-ary* structures in juxtaposition, rather than of two molecules with necessary identical structures... A general argument regarding compli-mentariness may be given. Attractive forces between molecules vary inversely with a power of the distance, and maximum stability of a com-plex is achieved by bringing the molecules as close together as pos-sible... in order to achieve the maximum stability, the two molecules must have complementary surfaces, like die and coin...

All the work done in biology since 1940 has proved Pauling and Delbrück to be right. The DNA double helix with its complementary chains is the best example. If a simple like-to-like templating process does turn out to explain the PrP to PrP^{sc} transition it will have to use molecular interactions currently unknown to science. If it is true it will be truly heretical. So perhaps Delbrück's ghost will walk after all.

Many other features of TSE infections and of the agents that cause them remain to be explained. A good example is the very long incuba-tion period. There are parallels in other brain diseases, like subacute sclerosing panencephalitis (SSPE) a complication of measles with many apparent similarities to CJD. SSPE occurred before measles vac-cination became widely adopted, with an annual incidence of about one case per million, and an incubation period of around ten years. Patients developed dementia before they died. But there is no doubt that it was caused by a conventional virus. The other fundamental bio-logical difference was the vigorous immune response mounted by the host, which is absence in CJD. Even if a satisfactory and detailed explanation of these long incubation periods is not yet to hand, which is not, (even for SSPE) at least it is known why prions do not stimu-late an immune response. It is because the body does not see its own proteins—including PrP—as foreign, and because the differences between PrP and PrP^{sc}, big as they are, are not great enough to cross this boundary. Another fundamental property of the infectious agent is its resistance to heat. The protein only prion hypothesis implies that PrP^{sc} must survive boiling and autoclaving. There are protein parallels for this in the bacteria which live in hot springs near volcanos and in places like Yellowstone Park. Does PrP^{sc} look like their heat resistant enzymes? Not particularly. They are full of alpha helices. But estab-lishment of the rules that determine protein heat resistance is still in its early stages. So it is not possible to give a molecular explanation of TSE heat resistance at this time.

The tenacity with which the protein only prion majority and the sceptical 'virino' minority hold their views is strong. An attempt at reconciliation by Charlie Weissman through an article in *Nature* proposing that PrP^{sc} could cause CJD on its own, but that a virino-like nucleic acid was responsible for strain differences, does not yet seem to have achieved its aim, despite it being one of the few occasions when humour was allowed to accompany a scientific paper. It opened with this story:

> Two men asked a rabbi to settle their dispute. After listening to the first man's presentation the rabbi announced, 'I find that you are right.' The second man, dismayed, protested, 'But rabbi, you haven't heard my side yet', and went on to present his case. The rabbi said, 'Now that I have heard you, I find that you are also right.' The rabbi's assistant interjected, 'But rabbi, they can't both be right', to which the rabbi responded, 'You are right, too'.

Whether 'virinos' exist or not, and whether PrP^{sc} acts alone or is helped by other molecules, it is clear that all the work directed towards establishing what TSEs are and how they interact with their hosts demonstrates that a good deal is known about their basic properties. Thus there is no controversy about their resistance to heat and chemicals, about their long incubation periods, and about the need to be prepared for the unexpected when considering interactions between a particular strain and a particular type of host. Their capacity to cause nasty iatrogenic surprises has been well demonstrated. We will return to these topics when we consider how BSE was handled. They were already very well known and abundantly documented before the first case was diagnosed.

There was a 1930s Southern Railway advertising slogan that went 'live in Surrey and be free from worry, live in Kent and be content'. The triumph of hope over experience, one might say. But important events happened in these counties at the beginning of BSE in the mid-1980s. If they had gone well, the slogan might have applied, at least for the human–bovine relationship. They did not.

Chapter 10

BSE

The Central Veterinary Laboratory (CVL) is located near Weybridge, in Surrey. In the 1980s it was part of MAFF and provided a diagnostic and consulting service for farm veterinarians and for the State Veterinary Service. Rare and difficult animal health problems from all over Britain were referred to it. It also provided advice to policymakers. In November and December 1986 the brains of three cows from Plurenden Manor Farm in Kent were sent to its Pathology Department. The first cow to fall ill at the farm with unusual symptoms had been seen by a private vet in April 1985. It was a Friesian/Holstein that had become aggressive and developed problems with coordination. Four other animals had fallen ill with the same pattern of disease during 1986. A senior CVL pathologist, Gerald Wells, examined the brains. His report and his diagnosis was clear. They all had a 'multifocal spongy transformation of the brain parenchyma and a degeneration of neurons, principally large neurons, in the brain stem'. He concluded that 'compared with most other animal disorders, the changes most closely related to scrapie, but there were

Figure 10.1   Uncoordinated and unhappy because of BSE.

subtle differences'. Rightly, he pointed out that 'it was not possible to rule out other degenerative condition, or entirely discount metabolic or toxic causes'. But there seemed to be little doubt that the illness was a TSE. In December another cow brain was found to have spongiform changes. This had been sent to CVL from Bristol on the 11th of the month (Figure 10.1).

On 19 December, the head of the CVL Pathology Department, Ray Bradley, sent a minute describing these findings to the Laboratory Director, William Watson, and its Director of Research, Brian Shreeve. He advised keeping an open mind until more information had been gathered because the principal lesions could be non-specific. But he also said that

> if the disease turned out to be scrapie it would have severe repercussions to the export trade and possible also for humans if for example it was discovered that humans with spongiform encephalopathies had close association with the cattle. It is for these reasons I have classified this document confidential...At present I would recommend playing it low key because a simple explanation may be forthcoming as a result of current investigations which will allay fears...You may also find the information valuable for defence of the CVL in a political sense.

Bradley's advice to keep an open mind until more information had come in was very reasonable at the time he wrote it, and his prediction of possible troubles to come regarding exports and human health was prescient. But his minute also reveals aspects of the mind-set at CVL that was to contribute negatively to the handling of the BSE crises. Clearly the members of the official veterinary world—all the CVL staff were civil servants—operated under very different constraints from their medical counterparts. It is hard to imagine the head of a hospital or university medical school pathology department sending a minute to his medical director or dean saying that his department had probably discovered a new disease but that the information should be kept quiet because of economic reasons or because quarantine barriers might be erected by foreign countries. And, it was not just CVL that felt it had to be careful. When William Watson rang up the government's Chief Veterinary Officer (CVO) to tell him about Ray Bradley's minute the CVO was seen 'walking down the passage with steam coming out of his ears'. He was worried about the economic implications.

In February another spongiform case in a cow was identified. This brain had come from Truro, in the South West of England. Evidence was also being found by electron microscopy that the new disease was indeed a TSE. The time had clearly come to consider whether the veterinary community should be told, in particular to encourage the submission of more material to improve the chances of finding out exactly what was going on. An in-house publication called *Vision* went to all the Veterinary Investigation Centres (part of the State Veterinary Service) and was an obvious choice for directed low-key publicity. Gerald Wells was unhappy, however. He thought a statement would be premature and that it would be 'damaging to scientific achievements at this laboratory'. Clearly he was worried about the possibility that CVL might not get the full credit for discovering this new disease. Scientists throughout history have been obsessed with establishing priority for the discoveries they make, and disputes about who first made a discovery are so common that the history of science is never a dull subject. Wells's concern was sharpened because a recent inspection of CVL had been somewhat critical and because low morale there was a problem. But Wells's worry about priority was not the only reason why the publication in *Vision* did not proceed. A memorandum from Ray Bradley to CVL colleagues gave the game away. In it he said 'we should consider what to do re the Scottish VICs (Veterinary Investigation Centres) who normally send neuro tissue to the Moredun. Some direction as to the restrictions and intentions as to future publications should be given'. *Vision* was circulated in Scotland. Clearly CVL wanted monopoly control over the new disease and any pathological material from dead animals. It solved its problem by circulating an article by Wells describing symptoms and pathology to Superintending Veterinary Investigation Officers in England and Wales on 8 June 1987. It was headed 'urgent' and 'in confidence'. It also included some directions. Among other things it said

> at this stage VI staff should not consult workers at Research Institutes or University Departments...A coordinated VIS/CVL publication on this subject is proposed. All statements on publication, or discussion at meetings MUST BE CLEARED by respective Directors of Services.

So, apart from a ten-minute presentation by Gerald Wells at the tail-end of a closed joint invitation-only meeting of the Medical and Veterinary Research Clubs on 29 May, no other official release of information was made. An attempt by Roger Hancock, a government vet in the South West of England to publish a letter in the *Veterinary Record* briefly describing the case from there was firmly quashed. Bernard Williams, Head of the Veterinary Investigation Service wrote to him

> further to our telephone conversation this morning, I am now confirming that the letter to the *Veterinary Record* which I cleared earlier in the week should not be published. I explained to you that this condition had been discussed by the CVO and the Director of CVL, and because of possible effects on exports and the political implications it had been decided that, at this stage, no account should be published. No doubt there will be an opportunity for your case to be published in due course.

Publication was not the only thing that was being firmly controlled. Richard Kimberlin, a member of Alan Dickinson's Edinburgh Neuropathogenesis Unit (NPU) had been at the Medical and Veterinary Research Club meeting and he had raised the issue of NPU involvement. This was also quashed—at least for the time being. Meanwhile, Gerald Wells had been preparing the draft of a paper for the *Veterinary Record*. The saga of whether the word 'scrapie' should appear in it captures the whole essence of the BSE crisis. It reveals the official mind at work in its full glory. It also set the scene for ten years of a seemingly proportionate but scientifically misguided and wrong response to it.

In mid-June 1987 Wells circulated a draft of his paper. The first half described the new disease. The second compared it to other TSE's, including scrapie. Wells was told that the CVO would not sanction it beyond 'the line on page 3'—the start of the second half. He replied that this was unacceptable to him and some of his colleagues. He would not be party to the modification of a scientific article for political purposes. He proposed a revision which would extend the discussion to 'a less biased view to the transmissible spongiforms'. But a minute from Ray Bradley to Wells confirmed that approval for pub-lication had been withheld for 'veterinary political reasons'. This line could not be held for long, however. Pressure was coming from State vets in the field. J. Gallagher from the Starcross VIC in Devon wanted to publish a short note. He put it nicely:

> I feel very strongly... that now with the delay in the publication in *The Record*, some note must be put in... if we are to retain creditability as a service, charged with the important functions of disease surveillance and the gathering of intelligence on new diseases, rather than suffer the embarrassment of what appears to be a total suppression of all informa-tion... while both numbers of practitioners and farmers know well of the occurrence of this condition. Our embarrassment will be consider-able when, as will inevitably occur, some other source will claim the surveillance trophy by reporting and detailing the condition first.

Any doubts that the new disease was significantly distinct from other TSEs—particularly scrapie—were dispelled through discussion with the NPU, which by the end of July was being more closely involved. The new condition had now been recognized in ten herds with nine more under investigation and 46 probable cases had been identified. The CVO finally gave in to arguments that it would be wrong not to discuss scrapie in a paper describing the new disease in early August. He asked that a paragraph he added to Wells' draft. 'It should be emphasized that at the present the etiological basis of the disorder recognized in England remains unknown and no connection with encephalopathies in other species has been established.' The paper was sent to the *Veterinary Record* on 17 August. It appeared on the last day of October, by which time there were 120 suspected and 29 confirmed cases.

It seems clear that the CVO's determination to censor out the word 'scrapie', and his successful attempt to cause the October *Veterinary Record* paper to distance the new bovine TSE from it, stemmed from a concern that the new disease was indeed scrapie in cows. It had become the lead hypothesis. In fact there did not seem to be any other. It was accepted as the idea to test by the epidemiologists. John Wilesmith headed the very small CVL Epidemiology department. By December 1987 and by dint of brilliant, almost single-handed work, he had become fairly sure that meat and bone meal (MBM) was the cause of the problem. The difficulty with this conclusion was that animals had been fed on it for a long time. Its use went back to the 1920s. Dead sheep had also been going into it for years, and scrapie was a very old disease. So to accommodate the notion that cows were now catching scrapie, as a new disease, there had to be a relatively recent change somewhere. Making some reasonable assumptions he worked out that it was probable that exposure of cattle had begun in the winter of 1981-2 and that the risk of effective exposure was greater for calves than for adults. Perhaps calves were more susceptible, or perhaps they had had a greater degree of exposure, or maybe both. Confirmation that MBM was a likely culprit—and that it was more likely than not that infection came from sheep—came from a zoo in Hampshire. In July 1986 a nyala had died from a spongiform encephalopathy and a year alter a gemsbok had similarly succumbed. The zoo had been using a pelleted antelope diet containing MBM that had come from Bockley Mills in Exeter, a plant that made 200 tonnes per week of sheep-derived MBM using temperatures that were not hot enough to kill the scrapie agent. In March 1988 the CVO set up a Task Force to look at the changes that had occurred in rendering—the cooking, extracting, and drying processes used in MBM manufacture. It reported rapidly after visiting 11 renderers and getting data from another 16 out of the 58 UK plants. Its findings fitted the scrapie hypothesis very nicely. Since 1980, to achieve economics of scale, there had been a drift from batch cooking to continuous rendering. It was thought that this would result in lower processing temperatures. Extraction of tallow (fatty material) by benzene-like organic solvents had almost disappeared and some low temperature renderers had started operation. The sheep population had gone up from 22 to 35 million, so more sheep material was being processed. More heads were being rendered and more casualty and condemned animals were being processed because knackers were going out of business. The numbers of scrapie cases had gone up, and the number of renderers had fallen from 200 to 58, meaning that pools of processed material were larger.

However, the leader of the Task Force, Peter Smith, made two important points in the conclusions of his report. Unfortunately they were soon lost sight of, with very unfortunate consequences. He said 'scrapie contaminated material has probably always entered and survived the rendering process but only in small amounts prior to 1980' and 'It is not possible to say what the critical limit is for agent relative to final feed concentration necessary to cause infection.' The obvious question posed by the first point is if BSE was nothing more than scrapie in cows why had it not occurred before? The second could be answered by experimentation; we will return to it later.

Wilesmith responded to the first question by reasonably hypothesizing that BSE had appeared because of the increase in the amount of scrapie agent in MBM, and that each case was directly infected by it—in other words, that there was no recycling of infectivity and no cannibalistic spread even although cattle went into MBM. This was very comforting for human health as well as for the prospects of controlling the disease in cattle. If BSE was nothing more than ordinary scrapie, it was safe, because people had been eating scrapied sheep for hundreds of years without apparent harm. Australians and New Zealanders had the same incidence of CJD as the British, despite their sheep being scrapie-free. And if large amounts of scrapie in MBM were needed to infect cattle, removing MBM from their diet would be a straightforward and effective control measure.

Epidemiology is an essential tool in the investigation of outbreaks of infection. From John Snow's time it has been a mainstay of public health practitioners. But it is insufficient on its own. It is always much better if it is used alongside biological information drawn from bio-chemistry, immunology, and genetics, just as crime investigations need evidence not only from witnesses but from fingerprints, tyre marks, and DNA as well. One can only speculate about what would have happened if Wilesmith's scrapie hypothesis had been tested by Dickinson in Edinburgh. But he was never formally told or asked about BSE. He had his hands full with other things in any case. As BSE was emerging in 1986 and 1987 he was fighting a battle to save his laboratory as an independent entity. In 1980 he had been appointed Director of the newly established NPU in Edinburgh. This was jointly funded by the Agriculture and Medical Research Councils, an unusual arrangement but one which took account of the hope that work on scrapie and other TSEs might shed light on the causes of human diseases like multiple sclerosis. But the Unit was being established at the beginning of the Thatcher era. The omens surrounding its birth were not good. Cuts were made to its budget even before it started work. Under Thatcher's government a new era had begun for agricultural research. The bad times had come. In essence, scientifically the subject was deemed to have had its day. Farmers had already got enough government money and they were producing mountains of food that nobody wanted. Industry should fund original work. The taxpayer should be let off the hook and public funding reduced. Perversely, as a big customer for the research MAFF was happy to oblige. Scientists responded by reorganizing themselves. Bigger groups were the order of the day. They would have lower administrative costs and so it could be claimed that money would be saved. 'Critical mass' was a concept much bandied about. Merging laboratories was seen to be a good thing. This is what Dickinson resisted. It was proposed that his laboratory should amalgamate with others to form the Institute of Animal Health. Dickinson would lose his Directorial authority and come under control of those who knew almost nothing about scrapie and TSEs. He dug his heels in. But he lost. He took early retirement towards the end of 1987 'weary and sad' for his NPU colleagues. Maybe he was right to walk away.

I went through similar battles in Aberdeen. The exercise there was called laboratory rationalisation—'lab rat' for short. It was very unpleasant. One of my colleagues termed the sessions at which pressure was applied on staff to change employers 'mangling meetings'. It was the second nastiest battle I fought in my whole professional career. This is not the place to tell the full story. Suffice it to say that I won the nastiest battle but in lab-rat the accountants won. The only consolation was that elsewhere things were worse. In fact it is no consolation at all because events like these played a big part in the sad story of the decline of academic medical microbiology in Britain. Currently it is close to extinction.

Even with his other pre-occupations Dickinson's views would have been worth seeking, because his experience of CJD transmission to humans was unique. He was the first person in the world to draw attention to the possibility that the disease could be spread through the use of human growth hormone extracted from pituitary glands collected from dead bodies undergoing post-mortem examinations. Not only that, but he proposed practical ways forward to address the problem, and carried out those that were in his power. Without doubt, if asked he would also had been able to say why, in spite of his warning, the story did not have a happy ending, and why its tragic conclusion carried important lessons for those conducting risk assessments of BSE.

Treatment of children who were failing to grow properly because they did not make enough growth hormone started in 1969. Growth hormone was extracted from human pituitary glands and purified by relatively simple processes. Until 1977 glands were collected and processed by the Medical Research Council because for most of this time the programme was run as a clinical trial. It was so successful in its aims that it was then taken over as a therapeutic programme by the NHS.

The pituitary gland is about the size of a big pea and is located underneath the brain. Lots of glands were needed for the preparation of human growth hormone, but their collection presented no particular difficulties. Examining the brain is part of the normal post-mortem routine. Removing it involved quite a bit of work—the skull cap has to be taken off using an electric reciprocating circular saw after folding the scalp down over the face and hooking it under the chin—but dissecting out the pituitary takes only a minute or two. I saw them being regularly collected at PMs for the growth hormone trial when I was a medical student in the early 1960s. The PM is not only a brilliant learning experience because it directly reveals medical failure as well as success, but is better than any textbook because of the surprises it regularly reveals. Not quite entertainment—but better than most detective novels and with a full house of auditory and olfactory side effects. For these reasons I often went to PMs in Lancaster during the holidays. Mr Pyle, 'Ernie' as he was called by some after the famous American war correspondent, was the mortuary attendant. He was not well paid. He showed films in private houses as an out-of-hours second job. Like most other attendants he collected pituitaries assiduously, and as in other mortuaries across the United Kingdom the MRC paid 20p a gland to cover costs and provide a fee for the attendant, a welcome addition to a wage of less than £8 per week. The glands were popped into a stoppered jam jar half-full of acetone kept on the window sill. When enough had been accumulated they were sent to Cambridge for processing. Attendants were not given precise instruction about the suitability of cases for gland collection. They used their common sense. Badly decomposed bodies were not used, for example. At this time about 65,000 glands were being collected every year in the United Kingdom.

On 5 October 1976 Dickinson rang the MRC. It had just occurred to him that there was a possibility that a scrapie-like agent could be transmitted in human growth hormone preparations. He was sure that such an agent could get through the extraction process. Storing scrapie material in alcohol—in practice no different from storing pituitaries in acetone—was a good way to preserve infectivity. Dickinson's warning was followed up by Dr Barbara Rashbass at the MRC. She rang him up and he offered to look at the protocols of the two different methods that were used to make the hormone to see if there was a step that would definitely exclude scrapie, and if not, to test them in a dummy run with material spiked with scrapie to see if it survived. The protocols were sent to him after a meeting at the MRC in December had discussed his concerns, and he replied on 22 February 1977 with the conclusions that he and Richard Kimberlin had reached jointly.

> In summary we think that there are reasonable grounds for optimism that the HGH samples prepared by *Dr Hartree's protocol would not be infective*, although you will see that many assumptions have had to be made. We would be somewhat *less confident for preparations using Dr Lowry's* protocol. However even if the fractionations did remove any infectivity, this would still leave the potentially serious problem of the early stages of extraction when the crude material would be a lab. hazard if infected tissue had been included in the batch. We can give advice on decontamination techniques if required but stress that infectivity can withstand many standard sterilisation methods.
>
> We suggest that four points should be considered by your committee.
>
> First, because of all the uncertainties in our assessment, it may be advisable to subject high titre scrapie mouse brain to the two protocols and test the end-products for infectivity. Second, an attempt should be made to assess the likelihood of unintentionally including a gland from a case of Creutzfeld–Jakob Disease, taking into account the frequency

at which cases are recognised clinically and also the fact that a high CNS titre can be reached preclinically. Third, it will probably be prudent to exclude glands from cases with dementia. Fourth, it may be advisable to keep reference samples from each batch which could, if necessary, be used to check for infectivity. The most appropriate and economical sample would be the first discarded fraction in each protocol.

So far, things had moved as fast as they could. But it was the last time that they did. Inertia supervened. Part of the reason was that Dickinson's warning had coincided with the move of the programme from the MRC to the Department of Health, a long-drawn-out and piecemeal process. Another was that two advisory committees ran the programme, an MRC Steering Committee which looked after hormone quality control and a Health Service Committee responsible for patient selection and treatment regimes. These arrangements created cracks for CJD worries to fall into. There was a shared responsibility for safety which meant that neither committee did its job properly. Eventually, at the end of November 1977 the MRC sought the advice of medical virologists. Peter Wildy, Professor of Pathology at Cambridge, replied on 1 December:

> The present situation is that the agents of Scrapie, Kuru, Transmissible Mink Encephalopathy and Creutzfeld Jakob disease present many similarities ... we are all forced to adopt the proposition (whether it be true or not) that all four types of agents will turn out to have similar properties ...
>
> We are presented with a set of imponderables that preclude any rational appreciation of the hazards involved.
>
> First we do not know for sure that scrapie-like agents are necessarily like scrapie in their physiochemical properties. Erring on the safe side, we must presume that they are.
>
> Secondly, we do know the Creutzfeld–Jacob disease is fairly uncommon (thought it may be commoner than we think) ...
>
> We are therefore left in the position that an agent that may be resistant to inactivation by many agents may be commoner than we expect. This could in the long run (though we don't know how long we must wait) give rise to a progressive or accelerated disease ...
>
> Now that there is definite evidence that the CJ agent will go in animals the resistance of the agent can be measured directly. This will take some months to do.
>
> At the same time it would seem to me wise to get Dickinson to check out a range of scrapie agents (there are many distinguishably different ones) to see how they would survive the treatments used in the extraction of GH. This will also take months but the sooner it is put in hand the better.
>
> In the meantime I am sure that the work on GH has to go forward and any clinical use of this hormone must be regarded as a risk, albeit

incalculable. We are in the uncomfortable position of suspecting the worst but not knowing how bad the worst is. Any clinician who uses GH must be made aware of the gruesome possibilities and their imponderable probabilities. It is the clinician who must take ultimate responsibility for his patients, but it is up to the Steering Committee to ensure that he understands the true position.

I don't know if this helps in the least—probably the reverse. But if you want to be more specific I shall be glad to help.

Despite the use of the words like 'gruesome', it was not until May 1979 that Dickinson was notified by the MRC that it was giving him a grant to carry out the scrapie 'spiking' test that he had recommended more than two years before. He was only able to test one protocol because of the shortage of facilities. Being scrapie, the experiments took years more. He finally concluded in November 1982 that a particular step in the Lowry process made the product safe.

Dickinson's other 1977 recommendations also had unhappy fates. Excluding glands from cases with dementia was not implemented until 1980. Reference samples were never kept. Worst of all, no attempt was ever made to assess the likelihood that a gland from a case of CJD or someone incubating the disease could be included in the pooled batches used to make a hormone preparation. Calculations done since have shown that although CJD is a rare disease in the population as whole, when account is taken of the age of post-mortem pituitary donors (most were elderly, the group in which CJD is commonest) and the practice of pooling many hundreds or even thousands of glands to make a single hormone batch, the probability of contamination was far from remote.

The first British growth hormone recipient to get CJD fell ill with it in 1984. In 1964 aged 2 she had a craniopharyngioma—a rare pituitary tumour—removed, and from 1972 to 1976 she had received twice weekly hormone injections at the Great Ormond Street Growth Hormone Clinic. When her illness began she was a secretary in a bank. It started with clumsiness, and unsteadiness when walking. Her behaviour became increasingly childish, demanding, and disinhibited, and in October 1984 she was referred to a psychiatric day clinic. By this time she could not stand unaided. By December she had become mute and unresponsive, and she died on 16 February 1985, aged 22. The natural hormone programme was stopped on 8 of May following the deaths of three recipients in the United States in April. By this time growth hormone produced by genetic engineering had become available and it was possible to switch the patients whose lives would be put at risk if treatment stopped suddenly onto the new product immediately. But of course by then the damage had been done. Nearly 2000 children had been treated in the United Kingdom with hormone extracted from the pituitaries of about 940,000 cadavers. So far 38 have died or are suffering from CJD. Its spread by growth hormone showed that at that time 139 cases had occurred worldwide; 74 in France, 35 in the United Kingdom, 22 in the United States, and 5 in New Zealand. The median incubation period of these cases was 12 years, ranging in individual cases from 5 to 30 years.

It is a curious, striking, and sad coincidence that at exactly the time that the first UK growth hormone CJD case was falling ill, the first definitive BSE cases in cattle were also developing their disease. The first animal to have a pathologically confirmed diagnosis belonged to Peter Stent, of Pitsham Farm in Sussex. A group of animals had developed coordination problems and nervousness in the summer of 1984. One of them, cow 142, was sent to CVL in September for diagnosis. It eventually turned out to have BSE. For good reasons the diagnosis was not made until 1987, when Gerald Wells mounted a retrospective analysis of brains stored at CVL to find out exactly when BSE had started. However remarkable this coincidence of onset of iatrogenic CJD and BSE it was nothing more than that. There can be no doubt that the CJD agent that infected the growth hormone recipients was of human origin.

Nobody had seriously suggested the BSE came from people. But where had it come from? In the early days Wilesmith's scrapie hypothesis was very compelling. However, as time went on, and more evidence came in some of its features became weaker and weaker, and it is no longer considered to be correct in its original form. The notion that BSE was unmodified scrapie and that it was caused by cattle eating rendered scrapied sheep material predicted that the distribution of different scrapie strains in cattle would mirror that in sheep. At the very least, multiple strains would be found, as in the sheep population. This turned out not to be the case. All BSE agents so far examined have been of exactly the same type—one that has never yet been found in sheep. The original hypothesis suggested that scrapie had spilled over into cattle not because it was efficient at infecting them, but the opposite. According to the hypothesis it had happened because there was more infectivity coming through due to changes in rendering and because more sheep brains and more sheep with scrapie were being processed. This implied that meat and bone meal would not be particularly infectious. It needed a lot of scrapied sheep in it to be dangerous. After all, it was known that no rendering processes, old or new, killed off the scrapie agent completely, and that scrapie had been reasonably common for hundreds of years. So meat and bone meal had probably had scrapie in it for many years. But meat and bone meal turned out to be remarkably infectious. Feeding as little as 1 g of BSE brain to an animal could kill.

The Wilesmith scrapie hypothesis relied heavily on the concept of the species barrier—that it was much more difficult to an infectious agent to infect a foreign species than the one in which had just been growing. It explained the recent jump from sheep to cow by postulating that this had happened because of a big increase in the dose of scrapie being ate by the cows. But it was still a puzzle why this had never happened before. Scrapie is unique among TSEs in its ability to spread from live sheep to live sheep in the field as though it was a virus or a bacterium. How this happens is not clear, but there is no doubt that it can persist in the environment in infectious form. Cows and sheep have been marching together for hundreds of years but scrapie in a cow had never been seen until BSE. If transmission was purely a dose effect it would not be unreasonable to expect it to have happened by chance many times before. But the Wilesmith hypothesis was not challenged in its early years. This allowed it to take a very firm hold. Most important of all was the endorsement it received from the Southwood Working Party.

Sir Richard Southwood's research specialty was insect ecology. In 1988 he was Professor of Zoology and Pro Vice-Chancellor of Oxford University, Chairman of the National Radiological Protection Board, and a Fellow of the Royal Society. He was right at the top of the tree of the British scientific establishment. His involvement was the first time a view had been sought from people outside government departments about the possible risks BSE posed to human health. Although MAFF had from time to time thought about this, it was only after Lord Montagu of Beaulieu wrote to the Minister of Agriculture, John MacGregor, on 4 December 1987 suggesting that BSE should be made notifiable with full compensation because of his concern that diseased animals were being slaughtered for human consumption that a paper was prepared for the minister. It went to him in mid-February 1988. The option it recommended was the one suggested by Lord Montagu. MacGregor's response was very cool. He had recently been Chief Secretary to the Treasury. Even though it was a measure to protect public health, it would set a precedent. He was particularly worried about rhizomania, a disease of sugar beet where diseased crops had been ordered to be destroyed without compensation. He suggested that the Chief Medical Officer, Sir Donald Acheson, be consulted. A letter was sent to Sir Donald on 3 of March. This was the first time that he had heard about BSE. He met with colleagues but nobody could say what the human health risk was. It was decided to set up a Working Party of experts. At the beginning of April he rang up Sir Richard to sound him out as to whether he would chair it. Sir Richard's response was 'why ever me?' Sir Donald replied 'because it is an ecological food chain problem and because I know you are an independent chap'. They had both served together on the Royal Commission on Environmental Pollution. Sir Donald had been particularly impressed by Sir Richard's skills as chairman. Although the press releases announcing its establishment called it an 'Expert Committee', none of its members were TSE specialists. In his evidence to the Phillips Inquiry Sir Richard said that he had agreed with Sir Donald that 'we should avoid those who were involved in the controversy surrounding the nature of the agent'. Officials had suggested that Richard Kimberlin should be a member. For a short time before retiring early he had been Alan Dickinson's acting replacement at the NPU. Sir Richard demurred. He wanted Anthony Epstein, a virologist, instead, and did not wish to have any experts who were 'almost too close to the front line to take the slightly broader view that we needed'. He said 'that another matter that was very much clouding the issue at the time was the Research Council's withdrawal of support from the unit at Edinburgh, which meant a number of the staff at Edinburgh were very unhappy and disillusioned'. Regarding Alan Dickinson, he said in oral evidence to the Inquiry that 'he had retired and, was not well, was suffering from actual psychological strain with all the traumas concerned with the funding of the institute he had brought up'. Sir Richard clarified this statement in a letter he wrote to the Inquiry a couple of days later 'I was seeking to pinpoint that the effect of these funding difficulties was such that Dr Dickinson himself took early retirement, being "battle weary", and therefore his personal involvement was lost at a critical time.'

Whatever the merits of breadth versus depth—and surely there is much to be said for both—for a scientific advisory committee to exclude real experts in the matter in hand from its membership was bizarre. But there was something very English about it. It was very firmly in the civil service mode of policy making by generalists—a body of individuals perceived to be dominated by public school men who had read dead languages at Oxford or Cambridge, people who were expert in ancient Greek poetry but who found simple statistics double Dutch. It could even be said that it might have drawn support from the particularly English tradition of the amateur naturalist. Entomology, Sir Richard's own specialty, is where this tradition can still be seen in action at its best. His membership of the editorial board of the *Entomologist's Monthly Magazine* points to a brilliant example. It is one of the few learned journals in the world that still publishes papers by professional researchers and by hobby collectors. That is not to say that its papers are of poor quality. Far from it. When I discovered a fly new to Britain it was the obvious place to publish my findings. Others have commented on this national tradition and BSE. The virologist Professor Fred Brown, who for a short time was a member of one of the scientific advisory committees that followed Southwood, commented in 1991 on the arrangement that had been set up to coordinate research: 'I would be hard pressed to devise a more inefficient way for studying the problem and ensuring practical messages come out of the work. It seems to follow the traditional English method for studying a problem by first forming a committee of amateurs...' He did not say this because he was anti-English. Far from it. At an International Virology Congress in Strasbourg I once heard him holding up a plenary session that he was chairing to give out the cricket scores in his strong Lancashire accent. Having said all this, there can be no doubting the eminence of the members of Sir Richard's Working Party, even if none of them had ever done any research on TSEs. They were top professionals in their own fields. Professor Anthony Epstein (later knighted) was famous for jointly discovering the Epstein–Barr virus in 1964. He had just retired from the headship of the Department of Pathology at the University of Bristol. Sir John (later Lord) Walton had just retired from the Chair of Neurology at Newcastle University and become head of a college at Oxford. He was also President of the General Medical Council. Dr Bill Martin was a vet who had just retired from the Directorship of the Moredun Research Institute in Edinburgh. But the task that faced them was a daunting one, because the remit that Sir Richard had negotiated was very broad: 'to advise on the implications of Bovine Spongiform Encephalopathy and matters relating thereto.'

Sir Richard's Working Party did not work alone. It had John Wilesmith, as an adviser, and a secretariat, half from MAFF (Alan Lawrence) and half from DHSS (Hilary Pickles). One of Lawrence's important roles was to keep his masters at MAFF informed about what the Working Party was doing. All of them wrote substantial parts of its Report.

Pickles lived in a boat on the Thames. She had a strong personality and the records show that she was combative. Descriptions of her activities attracted words like 'galvanize' and 'robust'. There was a minute from a MAFF official entitled 'Dr Pickles Strikes Again'. She had followed her Cambridge medical degree with research in pharmacology and a PhD. Before her involvement in BSE she had worked in the Department of Health on AIDS.

Once it met, the Working Party got off to a flying start. At its first meeting on 20 June it learned that animals with BSE were still going into the human food chain. Members were horrified and felt that they should stop it at once. Immediately after the meeting Sir Richard wrote to Derek Andrews, the MAFF Permanent Secretary, recommending that the carcases of ill animals should be destroyed by burning or by something equivalent. The Working Party also recommended that an expert group be set up to advise about BSE research, and that priority should be given to studies to find out whether BSE was transmitted from cow to calf and whether scrapie could be transmitted to cattle.

The recommendation that the carcases of sick animals should be incinerated was accepted and implemented without delay. At a stroke it removed a substantial mass of BSE infectivity from the human food chain and from meat and bone meal. So it was a very good thing. But it could, and should, have happened sooner. In the seven months between Lord Montagu raising his concerns with John MacGregor and the first Southwood meeting hundreds of animals with BSE had gone into the human food chain. If stopping this was purely a matter of scientific judgement it would have been right to wait for Southwood. But in essence it was down to common sense. Disgust is a basic human emotion. There is good evidence that through evolution—possibly as a mechanism to stop us eating dangerous microbes—our brains have developed regions that specialize in it. We learn to avoid faeces and foul smells when we are very young, and hold on to these feelings tenaciously until we die, go mad, or become demented. The initial gut feeling of MAFF officials that it was wrong to make pies and sausages out of clapped-out dairy cows (because it was these animals that were getting BSE) with holes in their brains was right, and should have prevailed without delay.

The pace of the Working Group's deliberations now slowed markedly. Its next meeting was in November. Richard Kimberlin attended it. He commented on what happened when giving oral evidence to the Phillips Inquiry.

> There were two real surprises. One was that this was actually only the second meeting that Southwood had had and here I was down there to give evidence, and then almost related to that was a major thrust of the meeting or at least a major preoccupation of that committee at the time was really on how to put together the final report. I am not for a minute suggesting that their minds were closed to other issues but there was a definite mind set about how they were going to do the report.
> ...I felt there was an awful lot of pressure there to get on with the report. So I ended up by actually not knowing quite what to do because I was expecting a question and answer session and it did not happen. Richard Southwood just said something along the lines: 'Well, Dr Kimberlin, before we have lunch do you have anything to add?' I had not really done anything, so why was I there?

As time went on the balance of power in the Secretariat shifted markedly to the DHSS. Hilary Pickles wrote minutes to 'galvanize' those responsible for the safety of medicines. She met with the Health and Safety Executive to ensure that occupational risks were being considered. She took over the remaining drafting (John Wilesmith wrote the chapters on epidemiology and 'The Future Course of the Disease'). Some at MAFF were unhappy. In February 1989 Alan Lawrence wrote an explanatory minute to Alistair Cruickshank, a MAFF Under-Secretary.

> The meeting discussed the timetable leading up to the publication and in view of their desire to complete the exercise quickly they considered which half of the Secretariat would be better placed to facilitate this. They may have raised this matter because they wanted to switch for tactical reasons or it may simply have been because they were aware that the preparation of the most recent draft (by MAFF) was not produced as promptly as it might have been. The result was that some of them had had little opportunity to read it before the meeting (on 16 December). The problem arose because, although special arrangements had been made to utilise the services of a Personal Secretary (with word processor facilities), the volume of work on salmonella at the time meant that she could not be deployed on the Working Party Report. In decided to ask Department of Health to take over the preparation of further re-drafts of the Report the Working Party may also have been swayed by the fact that the Department of Health half of the Secretariat is a grade 4, apparently with two Personal Secretaries, each with a word processor. In the circumstances it is possible that the decision to switch may have been taken simply on the basis of the facilities available to our respective Departments at the time.
>
> As a result of the changes it is perhaps not surprising that Sir Richard's communication with the Secretariat has largely been with the Department of Health side. However, the Department of Health have been keeping MAFF informed and, where appropriate, seeking our comments. But in a rapidly moving situation, where minor editorial changes can have a significant effect, there is always the possibility of being put at a disadvantage.

Pickles's own account of her take-over of drafting complements Lawrence's minute nicely:

> I...took over the final production of the report with Sir Richard's agreement.
>
> I understand Sir Richard had some concerns about delays in receiving minutes/drafts from MAFF and I judged that I would be able to do it more rapidly. I had better secretarial and word processing facilities, and this was the rationale given to MAFF for the change. I believe that Sir Richard also felt that my writing style was closer to that which was acceptable to the scientists on the Working Party. My note to CMO

after the last meeting of the working party hints at another issue. The concern was whether there were aspects in the report, and in particular the general conclusions just tabled by Sir Richard, that would be uncomfortable for MAFF and which, like some of the changes that had been suggested earlier, might be omitted in the final version of the report if the production were in MAFF's hands. The closure of offices over the Christmas break, and slowness in MAFF in reacting to this late amendment, enabled me to present MAFF with an agreed final text. Hence my hand delivery of copies of the draft report over Christmas to the (3) working party members who lived in Oxford.

At this time MAFF was giving much higher priority to Edwina Currie's TV statement about *Salmonella* in eggs than to BSE. In a curious coincidence, she resigned as Minister on the same day as the second Southwood meeting. The Working Party held its final meeting on 3 February. John MacGregor received its Report on 9 February and discussed it with Sir Richard and senior officials on 14 February. On the 22nd he sent a note about it to Margaret Thatcher and to MISC 138, the cabinet committee on food safety. The report was discussed at a full cabinet meeting on 23 February. The official News Release gave a fair and balanced summary of its conclusions and recommendations. It said:

> The report concludes that the risk of transmission of BSE to humans appears remote and it is therefore most unlikely that BSE will have any implications for human health. It points out that the related disease, scrapie in sheep, has been present in the UK for over 200 years and there has been no evidence of transmission to man. The Report also concludes that from present evidence it is likely that cattle will prove to be a 'dead-end' host for the disease agent.
>
> However, the Working Party point out that if their assessment proves incorrect the implications would be serious. The Working Party believe that the disease has developed because of feeding practices associated with modern agriculture. They suggest that, in the context of the adjustment of the agricultural policy of the EC in coming years, consideration should be given to changes in present methods. I have discussed this point with Sir Richard Southwood who has said that his Working Party are not recommending a ban on all use of meat and bonemeal in animal feeds, but are concerned that if they continue to be used they should be effectively sterilised.
>
> The Working Party have greatly welcomed the speed with which the Government acted to ban the use of animal feed rations in question and also to implement their interim recommendations. These measures include the compulsory slaughter of cattle suspected of having BSE and the removal from the food chain of their milk and carcases. The Working Party believe that the risks as at present perceived would not justify special labelling requirements for products containing either bovine brains

or spleen. The Report suggests, however, that manufacturers of baby foods should avoid the use of bovine thymus and offal. Sir Richard Southwood confirmed that the term 'offal' as it is used in the Report (as defined in the Regulations quoted) refers to brain, spinal cord, spleen and intestines (tripe). The Chief Medical Officer is satisfied that none of these, nor thymus, are used in the manufacture of baby food and advises that mothers ought not to feed these materials to infants of say under 18 months. As a precautionary measure I am however taking steps to ensure through secondary legislation that it will be illegal for anyone to sell baby food containing such products in future. With regard to other meats which people sometime refer to as offal, like kidney, liver and heart, the Chief Medical Officer advises that there is no need for concern. These are extremely nutritious foods which are beneficial.

As far as the Working Party's recommendations on animal feed are concerned we have already acted to ban the feeding of ruminant-derived material to ruminants; further action will be taken as necessary in the light of research work already under way on the heat-resistance of the agent and of further research which may be recommended by the Tyrrell Committee.

The Cabinet did not rubber-stamp the Report. In his evidence to the Phillips Inquiry Kenneth Clarke, then Minister of Health, said that

a member of the Cabinet . . . observed that the report appeared to be less precise than might have been expected of a report by eminent scientists on a matter of such moment; that its recommendations and suggestions were not based on precise and firm evidence; and that it was unsatisfactory that policy decisions should in effect be put in the hands of specialized experts.

Who this Cabinet member was Clarke does not say. As likely as not it was Margaret Thatcher. She was not happy about the Report. Neither were other Cabinet members. Attention focused on the offal in baby food recommendations—the 'matter of such moment'. It was said that it could seriously damage the baby food industry without being supported by any scientific evidence. However, at the end of the meeting, as Clarke says in his evidence,

the Prime Minister and the Cabinet were persuaded that it would not be sensible to refuse to implement any of the recommendations made in the Southwood Report. The Prime Minister summed up the strongest grounds in her judgement for adopting such an approach. She said that 'the Southwood Report was likely to raise considerable alarm, and its recommendations and suggestions were not based on precise and firm evidence. There were manifest difficulties in handling the situation, but it had to be borne in mind that the disease's recent crossing of the species barrier from sheep to cattle was a new phenomenon and it was not known whether a further crossing of a species boundary to humans was possible. In these circumstances the guiding principle was that the Government should be seen to act on properly qualified advice, and the Chief Medical Officer had a crucial role in this.'

Clearly, Thatcher had accepted the scrapie theory, but she had also spotted its central weaknesses as provider of good news, its argument about the species barrier. While postulating that scrapie had for the first time jumped to cows from sheep, it said that as far as humans were concerned the likelihood of transmission had not changed. It remained pretty well the same as for the old sheep disease. It is a pity that she did not follow up her concern. It was soundly based in logic as well as in science. Making prediction is about which species a *new* agent might infect is very difficult. With viruses, for example, the species barrier can be very strong, as in polio or smallpox, which only infect humans, or very weak, as with yellow fever which infects humans, monkeys and mosquitoes. Occasionally, as with some herpes viruses, a trivial infection in the natural host is lethal in another. A king of Greece died from such an infection after being bitten by a monkey. Changes in virulence and host range occur when microbes were grown in foreign hosts, and Dickinson's work had shown that the properties of scrapie changed after repeated cycles of growth in mice. But the unchanged scrapie theory held sway. This would not have mattered if continuing attention had been paid to the caveat inserted in the General Conclusion of the Report at the suggestion of Professor Epstein: 'if our assessment ... are incorrect ... the implication would be extremely serious.' But it was the conclusion that 'the risk of transmission of BSE to humans appears remote' that drove Government policy until the morning of 20 March 1996.

The baby food recommendation had very little scientific evidence behind it and the Cabinet was right to focus on it as a weakness. It would be reasonable to classify it if as another 'gut feeling', like the notion that old demented dairy cows with spongy brains were not fit to eat. But like that idea, good consequences come from it. Offal had been raised as an issue. Associating it with babies gave it a particularly strong flavour. So as an issue it did not go away. It was easy to understand why one did not have to be a vegetarian to feel unhappy about eating it, even as an adult. It was raised in Parliament by the Opposition five days after the Cabinet meeting. 'If, as appears likely to the Secretary of State for Health, BSE is a threat to humanity, why not ban the use of this offal for all human consumption? If according to the Minister of Agriculture, it is not a danger, why was it banned for babies?' The Woman's Farming Union called for a ban on brains for human consumption, and the Bacon and Meat Manufacturers Association advised its members to exclude pancreas, brain, intestines, spinal cord, and spleen from its products. But a key event in driving public policy was an initiative from the pet food industry. Richard Kimberlin had been commissioned by Pedigree Master Foods in July 1988 to advise on BSE. In May 1989 they invited MAFF officials to meet Kimberlin and learn about his recommendations. He had put offal into four risk categories. The highest was brain and spinal cord, and the next was parts of the intestine, lymph glands, spleen, and tonsil. This advice helped John MacGregor to decide that the best thing to do about the baby food recommendation was to extend the offal for babies ban to all human food. Administratively it was easier. It would deal with any sick animals that were slipping through the net. It would massively reduce any risk from infected animals that had not yet fallen ill (subclinical animals). None of the officials involved in developing this policy felt that it was necessary. But it would reassure the public. Sir Richard had to be asked about it diplomatically because it went further than his recommendations. Although he did not think that Kimberlin's views had changed the science, he could see the 'political necessity for action'. The ban was announced on 13 June 1989. The Pet Food Manufacturers Association immediately advised its members to exclude Kimberlin's risk offals from their products. Implementation for humans took another five months. There was a statutory requirement to consult. There were complex technical issues to consider. Tripe made from the fourth stomach of the cow contained some lymphoid tissue. Was there enough to justify a ban? A rule had to be devised to guide a decision. A pragmatic one was adopted. If the tissue could be seen with the naked eye, the product would be banned. Tripe escaped. The Bovine Offal (Prohibition) Regulations 1989 came into force on 13 November 1989. The bits of the animal that were banned were called specified bovine offals (SBOs).

So in the two years that had gone by since the identification of BSE as a TSE, humans had been protected from infection by removal of sick animals and the high-risk offals from the food chain. Implementation of these measures provided immediate and substantial reduction of risk. An equally important but longer term measure was a ban on feeding meat and bone meal (MBM) to ruminants to stop the transmission of BSE itself. The decision to do this was taken on 19 May 1988 and the Order making it an offence was signed by Agriculture Ministers in mid-June with a five-week period of grace before it took effect. It has been described as 'a spectacularly successful control measure...one of the notable success stories of global disease control'. But both this measure and the offal ban turned out to be very leaky. Evidence that a lot of infectious MBM was still being eaten by cattle after the ban started to appear in early 1991 when animals born after the ban (BABs) began to come down with BSE. A trickle soon became a flood. There were nearly 12,000 cases born after the ban in 1988 and over 12,000 born in 1989. Some were due to farmers feeding animals with old MBM bought before the ban. Some were due to accidental contamination of feeds. And some feedmills continued to put animal protein into cattle feed after the ban. Most of these sources of animal protein-containing feed ran out within a year. But BABs went on appearing in large numbers. A few were still being born in 1996. They were being infected by feeds that had been contaminated with MBM-containing products intended for pigs and poultry—which could legally still be fed MBM—during manufacture in feed mills. The essential problem here was that the idea that the infectious dose of MBM was high had taken a very firm hold among policy-makers. The original scrapie hypothesis still had a tight grip!

Stopping nervous system tissues going into human food was the most important aim of the offal ban. But the gaping hole through which it continued to travel was the portal provided by mechanically recovered meat. For good reasons this was a product unknown to the public. Bones from which meat had been removed by traditional methods were squeezed under high pressure to produce a slurry which, when freed of bone, went into low-grade sausages, burgers, and pies. The problem was that its main source was the spinal column—and that the removal of high risk spinal cord was not being done properly. All this was known in 1989. There was a meeting at MAFF to decide what to do. The note of the meeting spoke for itself. It recorded: 'The proposed ban on specified offals was itself a measure of extreme prudence, going beyond what Southwood recommended. Though some tissue would be contained in MRM it would be minimal and not present a significant risk. No action should be taken on MRM'.

Then came the cat. On 6 May 1990 the Agriculture and Health ministers were told that Max, a siamese cat, had been diagnosed at Bristol as suffering from a 'scrapie-like' spongiform encephalopathy. This event caused things to be done that should not have been done, and things to be left undone that should have been done. It should have been a signal that the time had come for the assumptions that underpinned the Southwood recommendations to be reviewed. If BSE was crossing species barriers that scrapie did not (attempts to infect cats with scrapie had not succeeded), maybe this indicated that it was not scrapie after all, and that the scrapie comfort factor was now very uncertain. But there was no review. Instead, reassurance about the safety of beef was order of the day. The Chief Veterinary Officer, Keith Meldrum (called 'Victor' by his critics because of a certain similarity of manner and appearance to the curmudgeonly TV character Victor

Meldrew) said that the cat was no cause for concern. The *Sun* didn't believe him. It compared BSE to the Black Death. The Meat and Livestock Commission said that there was no risk from eating beef. Sir Donald Acheson issued a press release:

> I have taken advice from leading scientific and medical experts in the field. I have checked with them again today. They have consistently advised me in the past that there is no scientific justification for not eating British beef and this continues to be their advice. I therefore have no hesitation in saying that beef can be eaten safely by everyone, both adults and children, including patients in hospital.

The highlight of the feline reassurance campaign was the Agriculture Minister's public attempt to feed his 4-year-old daughter Cordelia a beefburger. Because it was too hot her rejection of it was peremptory. Poor John Gummer. He had been challenged by a newspaper to show his confidence in British beef (Figure 10.2). As the Phillips Inquiry concluded, he had been put in a no-win situation. He lost in a big way—and on television.

As time went by more and more evidence came in indicating that BSE was very good at jumping the species barrier. By September 1994 57 cats had developed a TSE as had various exotic species including the Nyala, Gemsbok, Arabian oryx, Scimitar-horned oryx, Greater kudu, Eland, Moufflon, Puma, and Cheetah—all presumably through BSE-contaminated feeds. BSE had also been transmitted to the marmoset

Figure 10.2   The Gummers eating beef 'with confidence'.

by cerebral inoculation. But the official line remained that the risk to human health remained remote. Even if there was a slight shift from the original scrapie hypothesis because of these more recent findings, 'remoteness' was still the line to take because it was thought that the control measures that had been introduced more than took care of any potential problems.

The 'remoteness' policy had malignant effects. It encouraged a leisurely approach to the development of BSE controls and sloppiness in their implementation. It provided a firm foundation for vigorous defensive responses against those who disagreed with MAFF and DH policies. And it inhibited the preparation of contingency plans against things going wrong.

The best example of leisurely policy development is the bulls' eye saga. Although eyes are an extension of the brain, and could have been classified as specified offals, nobody ate them, so they escaped the November 1989 ban. They were still high risk materials, of course, and this raised the question about whether it was safe for schoolchildren to dissect them. It was discussed in general in MAFF. In February 1990 after a short consultation period the Scottish Education Department issued advice against using them for dissection. This stimulated Hilary Pickles to raise the issue with MAFF, and with Diana Ernaelsteen, Medical Adviser to the Department of Education and Science (DES). The matter was referred to the Spongiform Encephalopathy Advisory Committee (SEAC). In June they advised that eyes from animals older than 6 months should not be dissected in schools. In July Pickles advised Ernaelsteen about this and told her about the Scottish decision. The DES accepted responsibility for issuing guidance and on 28 August a draft submission recommending discontinuance was prepared for the Minister. But in September DES indicated a reluctance to act, and asked about pigs' and sheep eyes as alternatives. In October Ernaelsteen replied to say that sheep eyes were unsuitable but those from pigs, horses, and calves under 6 months were acceptable. A start was made on a second draft submission to the Minister. In January 1991 it was ready. In February Ernaelsteen asked why there was a delay in issuing advice. In late April 1991 a revised draft was circulated. In May MAFF said that it was content. But DES officials were busy with other things, and meant the third draft did not go out for cross-departmental approval until March/April 1992. DH was content and so was MAFF, who queried why things were taking so long. The Health and Safety Executive doubted that there were any problems. In May, however, Ernaelsteen had second thoughts. Perhaps the media would be interested. Perhaps the advice would have a negative effect on human growth hormone treatment. Issuing the advice now might cause people to ask 'why now? Why wasn't this issued before? What has changed?' In June DH asked why the delay? It repeated the question in September; DES replied that because of Ernaelsteen's advice it has low priority. In October DH said that the advice should be issued; flouting the advice from SEAC would be the first time that the Government had ignored a recommendation from this advisory committee. Within a couple of weeks DES replied to say that it would put the submission to ministers it went to the Minister of State on 29 October 1992 and the guidance recommending discontinuance was issues to English schools in mid-December, nearly two years after the Scots had received it. The Phillips Inquiry summarized all these goings on very neatly. It was 'the best being the enemy of the good'. BSE had tested the Government machine and found it wanting.

Nobody had ever suggested that bulls' eyes posed a major risk or constitute an important route of infection. The same could not be said for spinal cords. They were included in the offal ban. Although there is no doubt that more attention was paid to their removal after it came into force than before, what went on happening is as good as example of sloppiness as any other in the BSE story. Carcases are split down the backbone with a power saw. Although checking that the exposed cord is removed is a straightforward exercise, it needs care and attention to make sure that nothing is left. But no emphasis was put on this when the ban came into force. No part of the food chain was then (and still is) inspected more than slaughterhouses, with their resident meat inspectors and vets. The State Veterinary Service visited as well. It was as though every big restaurant or butchers shop had a resident environmental health officer rather than being visited every few months. But big chunks of cord still went into human food and MBM. How many we shall never know. But the problem was not resolved until the inspection regime had been radically reformed by taking it away from local authorities and establishing a national inspectorate, the Meat Hygiene Service. This change was proposed by John Gummer in July 1991. It had nothing to do with BSE. It ran into opposition as an over-regulatory measure so its establishment took a long time. It took over from local authorities on 1 April 1995. It found that on average four carcases in every thousand still had some cord. The inspection regime had failed. 'Remoteness' had encouraged sloppiness.

'Remoteness' was defended with vigour. A dramatic demonstration was the letter that Hilary Pickles wrote to the Chief Medical Officer at the Welsh Office, Deirdre Hine (later Dame Deirdre) responding to her concerns about BSE and human health. Dame Deirdre had been advised by Stephen Palmer, an epidemiologist working for the Public

Health Laboratory Service (PHLS) based in Cardiff that there was insufficient evidence to say categorically that there was 'no scientific justification to avoid eating British beef'. Her advisers were also worried about how well the specified offal ban was working, and were unhappy that no medical epidemiologists were investigating BSE. These were all legitimate, and as it turned out, very well founded concerns. The best way to describe what happened is to quote from Dame Deirdre's evidence to Phillips. She was being questioned by Mr Freeman, a Counsel for the Inquiry:

MR FREEMAN: You recognise this letter of 9 August 1990?

DAME DEIRDRE HINE: I do.

MR FREEMAN: Now, just going through the letter for those that have not seen it before, it says: 'I am unhappy that you are discussing matters with MAFF before discussing with us, particularly as you are questioning a study commissioned by the Department of Health.'

What do you think was wrong with discussing matters with MAFF?

DAME DEIRDRE HINE: Well, in fact we were not discussing matters with MAFF. I was discussing our concerns with my agriculture colleagues in the Welsh Office because clearly some of our concerns impinged on their area of responsibility and it was they, in their quite proper relationship with MAFF, who were then relaying those concerns to MAFF.

We, at the same time, were telling the Department of Health exactly what we were talking to our agriculture colleagues about, so I do not think that Dr Pickles' unhappiness was really justified.

Then after more discussion about technicalities:

MR FREEMAN: Then there are quite a lot of paragraphs after that: 'We are reviewing all the available published data (10,000 plus references) on spongiform encephalopathies.' 'Mr Davies raised the abattoir worker study...'

Then it says:

'Since Mr Davies' letter was copied to the Chief Medical Officer I have copied this also. Mr Davies' letter claims Welsh Office are not trying to embarrass colleagues here but I do think the way you have handled this is surprising.'

What did you think of that suggestion?

DAME DEIRDRE HINE: I was surprised that they thought the way we had handled it was surprising. I thought that we had handled it entirely properly in both Welsh Office Agriculture Department, communicating with their colleagues in MAFF, and ourselves making simultaneously our concerns, which actually we had previously expressed to the Department of Health, making those concerns known to them again.

MR FREEMAN: The next sentence says: 'Many of the comments suggest you have a keen interest in BSE and CJD but are in possession of only some of the available information. In these circumstances, surely it would be preferable to check out your ideas on the phone before they get formalised into an official viewpoint put to another Government Department.'

What did you think of that suggestion?

DAME DEIRDRE HINE: We did not have a keen interest. We had the interest of the people who had a responsibility to a population.

MR FREEMAN: Namely the population of Wales?

DAME DEIRDRE HINE: Yes. We were well aware that we were not in possession of the full facts, though we had tried very to obtain access to the Committee and to all the evidence, and we were getting very much generalised responses back, rather than specific responses.

However, I was, to a certain extent, understanding of the position of my colleagues in the Department of Health who were dealing with very difficult, very sensitive matters, on which we did not have the whole information and therefore we were asking awkward questions and we, of course, had the luxury of not having the responsibility for the policy lead. So to some extent I can understand the slight degree of irritation which is creeping into the correspondence.

MR FREEMAN: It goes on in the next paragraph: 'For my own part, I do not see there was a particular Welsh 'angle' to BSE/CJD, and am surprised you feel it necessary to put so much effort into challenging the views of colleagues at DH who are more senior, more experienced in the area, devote a higher proportion of their time to the topic, and have frequent access to the real experts in the field.'

What did you think of that?

DAME DEIRDRE HINE: All of that was true. They were more senior. In fact, I at Grade 3 was appreciably more junior to many of my colleagues in the Department of Health. They were more experienced, they did have a higher proportion of their time, and all that is absolutely true. However, I did not think that absolved me of responsibility for making known my concerns to them.

Only the more lurid parts of Hilary Pickles's letter have been reproduced here. No doubt they owed a good deal to the Pickles personality. In reading them it would be wise to take account of Edwina Currie's view. She found her 'rather frightening...extremely forward'. But the letter was an accurate reflection of Whitehall policy. Dame Deirdre never got Welsh representation on the Spongiform Encephalophy Advisory Committee. Her concern that no medical epidemiologists were official working on BSE went unheeded, particularly her wish that 'their involvement would be tantamount to admitting the possibility of a human health risk'. This was a great pity. Not only did her PHLS epidemiologist colleagues in Wales have enormous expertise in food-borne disease, they had 'squelched around abattoirs'.

The 'remoteness' policy met with a sudden end. It was killed by the SEAC on Saturday, 16 March 1996. Events at this time moved very fast. They showed that no thought had been given to the possibility that the policy might have been wrong.

As a scientist-dominated advisory committee SEAC was unique in the deference that the Government paid to its views. 'Waiting for SEAC' had been the policy-makers mantra ever since its first meeting on 1 May 1990. SEAC came out of Southwood. In his first report in June 1988 he recommended the establishment of a working party to advise on research. For the next four months MAFF and DH battled over its chairmanship. Sir Richard and DH favoured a virologist and proposed Sir Michael Stoker, a 70-year-old long-retired expert on tumour viruses. MAFF wanted a vet. They put up several names, but lost. At the beginning of November it was agreed to appoint David Tyrrell, Director of the Medical Research Council Common Cold Research Unit. The first meeting was in March 1989. His report went to the Government in June. They published it in January 1990. In April 1990 the Working Party was reconstituted as SEAC. SEAC met several times a year. In 1995 it had its 18th to 22nd meetings, and a new chairman took over in November, John (later Sir John) Pattison, another virologist. John Collinge, a Professor at St Mary's Hospital and the first active TSE researcher ever to be a member, joined in December. He had met with the Chief Medical Officer, Kenneth (later Sir Kenneth) Calman a couple of months before to tell him that he feared that cases of CJD in young people might be due to BSE transmission to humans. At the SEAC meeting on 5 January 1996 Collinge's concerns mounted. SEAC met again on 1 February. By now there were five cases. Collinge's worries—shared by Pattison—were minuted. But the notes about these meetings passed on to DH and MAFF did not accurately reflect their fears.

The Prime Minister, John Major had been briefed about the teenage cases in October and November and told that because CJD had been found in teenagers in other countries before the appearance of BSE it argued against a link. He was briefed again in early December because Sir Bernard Tomlinson, a retired neuropathologist had said on the BBC 'You and Yours' programme that he would not eat beefburgers. Major was told that these remarks were unnecessary and were unjustified—Tomlinson obviously did not know that the Government had announced on 28 November that bovine vertebrae were no longer going to go into MRM, or that BSE numbers had fallen dramatically in the last couple of years. At a Cabinet meeting a couple of days later it was agreed that yet again the time had come for a response to prevent public panic. A key message would be the assurance from Government advisers that there was no evidence that the disease could be transmitted to humans. But at the SEAC meeting on 8 March the view began to crystallize that the young CJD cases had probably been caused by exposure to BSE. Word about this went to MAFF and DH. They passed the buck back to SEAC. It held an emergency meeting on Saturday, 16 March. There were now nine confirmed and three suspect young CJD cases. SEAC concluded 'this is great cause for concern. On current data the most likely explanation at present is that these cases are linked to exposure to BSE before the introduction of the SBO ban in 1989'. On Monday, 18 March Douglas Hogg, the Agriculture Minister, discussed with his officials that there should be a ban on selling animals over 30 months old. The Prime Minister, John Major, and other Cabinet members became involved. At a Ministerial meeting on the morning of Tuesday, 19 March it was eventually decided—again—to wait for SEAC. They were pressed to have an early meeting. It took place at 4 PM. Some members were at a scientific meeting in Paris and contributed by phone. It adjourned late in the evening without reaching a conclusion but with a message that the Government needed advice by 10.30 the next morning. It met again at 8 AM with phone links to Edinburgh and Paris. The Cabinet met at 10.45 AM. At 3.31 PM Stephen Dorrell, the Minister of Health, rose in Parliament and announced that ten young people had contracted a new variant of CJD and that it was probable that they had caught BSE. He was followed at 4.17 PM by Douglas Hogg who among other things confirmed that the Government would follow SEAC's recommendation that carcasses from cattle over 30 months old must be deboned in specially licensed plants.

Within two weeks deboning had been abandoned. It had not reassured the public. The supermarkets declined to sell meat from animals older than 30 months. Deboning capacity was not enough to cope. SEAC had been pressurized into making policy on the hoof. It had not worked. There was no contingency plan.

Chapter 11

BSE—Why Things went Wrong

For many years 'remoteness' was acted on as if it was a scientific fact. Why? The brilliant analysis of the Polish microbiologist and immunologist Ludwik Fleck provides the answer. In his 1935 book, *Genesis and Development of a Scientific Fact*, he explained how science works as a collective enterprise. Discoveries are not made in a vacuum. They are deeply influenced by the training, experiences, attitudes, and assumptions of researchers. So is the way that they are presented and received. Fleck calls these things a 'thought style'. He mapped all those with an interest in a particular discovery in a 'thought collective'. Its structure is like a Russian doll. Specialized experts working on the problem are at the centre. They practice 'journal science'. It is provisional, raw, and cautious. General experts comprise the next layer. Fleck calls their view of the subject 'handbook science'. Thinking focuses on results that are judged by them to be important. Contradiction is resolved by negotiation; conflict by demagogy. Handbook science is nearly always a little out-of-date. Practitioners of journal and handbook science form the esoteric circle. The outermost and biggest layer of the collective is the exoteric circle. Its members are educated amateurs. Their understanding comes from specialized—and provisional—esoteric knowledge, but, Fleck said, 'owing to simplification, vividness, and absolute certainty, it appears secure, more rounded, and more firmly joined together. It shapes public opinion ... and reacts in turn upon the expert'. For BSE, Southwood and SEAC comprised the outer ring of the esoteric circle; exoterics were civil servants and government ministers. 'Remoteness' flourished during the simplification and the removal of caveats and qualifications that happened when information was transmitted from the esoteric to the exoteric circles. The influence of the exoteric circle on the esoteric reinforced it. Veterinarians and public health officials may have known little about TSEs and their iatrogenic tendencies but they all knew that scrapie was harmless for humans.

The 'thought style' of the esoteric circle is vigorously defended. Fleck put it dramatically: 'heretics who do not share this collective mood ... will be burned at the stake ... ' In essence this is what happened to Deirdre Hine and Richard Lacey, the Professor of Microbiology from Leeds whose predictions of a massive CJD outbreak received much media attention from 1990 onwards. Even if there were no *autos de fe*, neither they or other members of their own esoteric circle, public health doctors in Wales for Hine, and Stephen Dealler, a microbiologist who worked for Lacey, were allowed to join the BSE thought collective.

A severe chronic disease of the government machine also helped to perpetuate 'remoteness'. 'Departmentalism' has flourished ever since anyone can remember. It was given a boost in 1918 by the Haldane Committee on the Civil Service which supported the model of departments organised to provide a service rather than to serve clients—which would have required cross-cutting structures. Its main symptom is the 'turf war'. Departmentalism is harmful because it causes a ministry to resist changes beneficial to the public because they threatens its interests. Perhaps they would diminish the departmental budget, or its staff numbers, or its standing. 'Not invented here' is a powerful stimulus for a negative response to a change proposed by someone else. Departmentalism is brilliantly illustrated by Hilary Pickles's letter to Deirdre Hine. The bulls eye saga illustrated its malignant effects on joined-up policy making and implementation. A related feature that inhibited a review of the 'remoteness' policy was identified by the American Don K. Price in his classic *Government and Science*. He said that the UK civil service is

> a profoundly conservative force ... in the sense of looking on the government and its programme as a single coherent machine in which inconsistency cannot be permitted. Any novel idea is an inconsistency

that could cause temporary waste and disorder and inefficiency and would probably detract from the current program.

But the government machine is not always so well-oiled. The concept that organizations have rationally managed, precise, consistent and stable goals is challenged by the 'Garbage Can' model of decision making. Developed in the 1970s it emphasized the impact of organizational structures, the indifference of some decision makers and the bias of others, personal agendas, accidents of timing, conflicting interests, the complex interaction of interesting parties, and happenstance. The 'choice opportunity'—the meeting to produce decisions into which this mixture is dumped—is the garbage can. Scott Sagan said in his book *The Limits of Safety* said that the model has 'illuminated the hidden and more capricious aspects of organisational life'. The establishment and composition of scientific advisory committees on BSE owed a lot to the garbage can system. Battles over chairmanships and secretariats were hard fought because of departmentalism. The letter from Lord Montague of Beaulieu and the Chief Medical Officer and Sir Richard's joint membership of an unrelated Royal Commission played key roles in driving decisions rather than well-established rational procedures.

In Fleck's analysis the centre of a thought collective is occupied by specialized experts who are active researchers on the problem. But with the exception of John Wilesmith—who also had other very important jobs to do at Weybridge, like working out the epidemiology of tuberculosis in cattle—and Bob Will, whose main role was clinical surveillance, from 1988 to 1995 they were absent from the government's scientific advisory apparatus. Attempts to involve Public Health Laboratory Service epidemiologists were rebuffed; they were expert in human infectious diseases, including those transmitted from animals, but it was thought that using them would give a signal that 'remoteness' was being questioned. We will never know what the outcome would have been if TSE experts had joined the thought collective through membership of advisory committees. Nevertheless, it is a reasonable guess that the science of BSE would have moved faster, that its full range of unique properties would have been demonstrated and appreciated earlier, and that in consequence control measures would have been better and the government's surprise by vCJD mitigated.

The obvious experiment to test whether scrapie causes BSE is to feed or inject cattle with infectious material from sheep. It was recommended in a letter from Sir Richard Southwood to the Permanent Secretary of MAFF in June 1988 (although it was not specifically mentioned in his Working Party Report or in that of the

Tyrell Committee on research). But Sir Richard's recommendation was not acted on. Policy makers came up later with many excuses. The experiment had already been done in the United States. It would be expensive. There was a large number of scrapie strains in British sheep, so to test the possibility that only one of them caused BSE all would have to be examined, making it even more expensive. Because of uncertainty about whether the right strains of scrapie had been tested it would be difficult to interpret. It was as though nobody wanted to lift the foundation stone of the scrapie hypothesis in case there might be something unpleasant underneath—like evidence against it. The US experiments certainly gave it no support. The first one had been done in 1979. Three out of ten animals had fallen ill. Neither their illnesses nor the pathological changes in their brains resembled BSE.

The ELISA (Enzyme Linked ImmunoSorbent Assay) saga was another scientific development that went much more slowly than it should, or could. A test to detect ruminant protein in animal feed would have helped enforcers of the MBM ban enormously. An ELISA test would do this. The technology involved had been around for many years. It was not rocket science. MAFF decided to develop it in-house rather than by seeking external collaboration; this meant that they would retain the intellectual property rights. Work started in February 1989. Field tests in 1991 showed that its development was not going smoothly; most feeds produced a positive result even when they included no MBM. In 1993 the Worcester laboratory where the ELISA work was being done was relocated, leading to a suspension of its development. The test went into limited use in 1994, despite unresolved imperfections. It showed that MBM was still in cattle feed, even though it had been banned from it in 1988.

Despite the knowledge that experiments took a very long time because of very long incubation periods, they were often done very late. The initial results of experiments to find out how much infectious material would infect a cow with BSE (1 g) only became available in the summer of 1994. Some experiments have never been done. We do not know how many scrapie strains are present in UK sheep, or the full range of their properties. Whether one of them is like BSE, and spread to cattle, remains an open question because of this ignorance.

So BSE science could have been better managed. Were there helpful precedents? Fred Brown thought so. In 1991 he wrote to the Department of Health, MAFF, the Research Councils, and the Government Chief Scientist recommending that an 'Oppenheimer' should take charge of BSE science. J. Robert Oppenheimer had been appointed in the autumn of 1942 to direct Project Y, the Los Alamos programme to build the atom bomb. It completed its task in 27 months.

But wartime conditions gave advantages that no BSE supremo could ever have. Oppenheimer's unlimited dictatorial management powers were matched only by his lavish funding. But Brown was right to look to the atom bomb programme for lessons from history. In her outstanding account of science and government, 'Britain and Atomic Energy 1939-1945', Margaret Gowing described the MAUD Committee as 'one of the most effective scientific committees that had ever existed'. Both its origin and its name go back to Lise Meitner. Forced to leave Germany for Sweden, in 1938, she spent Christmas with her nephew, Otto Frisch, who was working with the Danish physicist Niels Bohr in Copenhagen. They discussed some curious results from Germany. It was evident that uranium had undergone fission. The theoretical implications were obvious. An atom bomb was possible. In the summer of 1939 Frisch moved to Birmingham. He worked with Rudolph Peierls, Professor of Applied Mathematics, who had left Germany in 1933. Being German they were excluded from war work, so they thought about nuclear fission instead. In March 1940 they produced a three-page typescript which showed that an atom bomb was technically possible, how destructive it would be, how much fissionable material would be needed, and how this material could be made. It said 'the bomb could probably not be used without killing large numbers of civilians, and this may make it unsuitable as a weapon for use by this country'. They had no ethical qualms about their work. To them its ideas were so obvious that there was no doubt that the Germans, with their strengths in physics, would already be well advanced down the same road. The Frisch–Peierls Memorandum was uncannily accurate. Gowing described it as: 'a remarkable example of scientific breadth and insight. . . . The two scientists had performed one of the most important and difficult tasks in science—they had asked the right questions . . . they had also answered them correctly from theory without any experimental aid'. The Memorandum rapidly reached Professor George Thomson, a physicist at Imperial College, London. A uranium sub-committee of the Committee for the Scientific Survey of Air Warfare was set up. In April 1940 it called itself the MAUD Committee for camouflage. The name came from a telegram sent by Lise Meitner at Bohr's request shortly after the Germans had overrun Denmark. 'Met Niels and Margrethe recently. Both well but unhappy about events. Please inform Cockcroft and Maud Ray Kent'. Cockcroft was a leading British physicist. But what did 'Maud Ray Kent' mean? Frisch and Peierls substituted 'i' for 'y' and produced the anagram 'radium taken'. Another suggestion was that it was said 'make Ur day nt' (make uranium day and night). Only after Bohr escaped to England in 1943 (in the bomb bay of a Mosquito aircraft) was it discovered that Maud Ray had once been a governess to his children. She lived in Kent. Fifteen months after its establishment MAUD delivered its reports. They were evidence based. Lots of research had been done. The rest is history.

Why was MAUD so successful? It ignored politics. It did not lobby ministers or civil servants. It was poorly placed in the Government (it was attached to the Ministry of Aircraft Production). It operated in secret. It had no follow-up powers. Many of its experts were foreigners who had to get official permission even to possess a bicycle. Paradoxically, this helped. Peierls had only just been naturalized and Frisch was still an enemy alien and it was not possible to put them on the main committee. So a Technical Sub-Committee was set up for them. This is where MAUD's important work was done. Most senior British physicists in the field were members. There was a full exchange of ideas and freedom of discussion. Gowing said that 'its members displayed in their work in the laboratories and their discussions in committee, qualities that may be simple in themselves but are more rare in combination—professional and administrative competence of a very high order allied with intensity and unity of purpose. Herein lay the MAUD success'.

The MAUD reports went from its sponsor, the Director of Scientific Research of the Ministry, to the Defence Services Panel of the War Cabinet Scientific Advisory Committee. This resembled Southwood in that its members were eminent scientists whose job was to take the broader view. Only one was a physicist; three were biologists and one a chemical engineer. Three were Nobel laureates. They met seven times in September 1941, taking evidence from witnesses; they recommended that the bomb project was of the highest importance. This reinforced a decision already taken. The Prime Minister's scientific adviser, Lord Cherwell, had followed MAUD closely and briefed him about its reports. Churchill was positive: 'Although personally I am quite content with the existing explosives, I feel we must not stand in the path of improvement . . .' Sir John Anderson, a member of the War Cabinet, was given responsibility for taking things forward. By chance, at one time he had done research on the chemistry of uranium.

The forces driving a wartime scientific advisory committee are special. But MAUD's inclusion of all major scientists working in the field, its facilitation of the exchange of ideas and information, and the testing of hypotheses it sponsored could be replicated at any time. BSE science suffered because they were not.

Sir John Anderson's background as a former uranium chemist made it easy for him to judge the arguments underpinning the bomb project. But even in the twenty-first century he remains the only scientist ever to hold the key government jobs of Home Secretary and Chancellor of the Exchequer. This is unlikely to change in the near future, even if its probability is going to be greater than that of future senior civil servants and government ministers previously being Governor of Bengal and having an air-raid shelter called after them, some of Anderson's other achievements. Rules must therefore be constructed to enable policy makers, the 'amateur' members of Fleck's exoteric circle, to assess the 'esoteric' advice received from scientists. How to test its validity is the question. Judge Harry Blackmun of the US Supreme Court provided a way forward in his 1993 'opinion' in the 'Daubert' case. Jason Daubert was born with serious birth defects. His parents and others sued Merrell Dow Pharmaceuticals alleging that they had been caused by the mothers taking Bendectin, an anti-nausea drug. The Supreme Court became involved because of disagreements about standard setting for expert evidence. Blackmun pointed out that there are

> important differences between the quest for truth in the courtroom and the quest for truth in the laboratory. Scientific conclusions are subject to perpetual revision. Law, on the other hand, must resolve disputes finally and quickly. The scientific project is advanced by broad and wide ranging consideration of a multitude of hypotheses, for those that are incorrect will eventually be shown to be so, and that in itself is an advance. Conjectures that are probably wrong are of little use, however, in the project of reaching a quick, final, and binding legal judgement.

In summary, he said that tests of scientific validity should include

> whether the theory or technique in question can be (and has been) tested, whether it has been subjected to peer review and publication, its known or potential error rate, and the existence and maintenance of standards controlling its operation, and whether it has attracted widespread acceptance within a relevant scientific community. The inquiry is a flexible one and its focus must be solely on principles and methodology, not on the conclusions that they generate.

The relevance of Blackmun's 'opinion' to scientific work on BSE is very great. It is easy to find instances where its principles were not followed, with bad consequences. The law wants quick answers. So did Sir Donald Acheson when he asked Sir Richard Southwood to consider the available data and give a considered view about the implications of BSE for human health. Sir Richard's role was the same as a judge who had heard evidence from expert witnesses. However, Blackmun reminds us that scientific conclusions are subject to perpetual review. This never happened to the Southwood conclusions, and 'remoteness' became set in stone although differences between scrapie and BSE became more and more evident, facts that cast increasing doubt on the hypothesis on which it rested.

Regarding the validity criteria, we have seen that there was a great reluctance to do experiments to *test the hypothesis* that scrapie was the same as BSE. There was an unreasonable delay in publishing papers in *peer-reviewed journals* when BSE was first discovered (for veterinary–political reasons!) For a long time wildly optimistic assumptions were made about the infectivity of feed containing MBM; a cross examination in court would have revealed that these came from guesses and that it would have been pointless to ask about *error rates* in measurements because none had been done. Major questions were raised in October 2001 about *quality control* at the Institute of Animal Health, a major contractor for TSE research in Britain, when checks on brain material collected from sheep with scrapie in the early 1990s and being tested for the BSE prion showed that it was all bovine. There was no sheep at all. During the early years *widespread acceptance within the relevant scientific community* was not tested as vigorously as it could have been because CVL kept a tight grip on research and research material.

Daubert and MAUD provide guidelines and models for the future as well as providing templates against which past scientific performance can be measured. Their usefulness is restricted to process, how things should be done. Judge Blackmun's 'opinion' is particularly helpful for policy makers because it is concise and written in careful English for lawyers. MAUD was successful because of the way it worked. But from its start it had another enormous advantage. The Frisch–Peierls memorandum contained a simple formula describing how much ^{235}U was needed to make a bomb, how it should be constructed, and how much energy would be released. All its parameters could be accurately measured. Frisch and Peierls's initial guess of the crucial one was only out by a factor of four. The degree of reliance that could be put on mathematical modelling was so great that the first ^{235}U bomb was used on Hiroshima on 6 August 1945 without prior testing. There is, however, no formula for BSE or for vCJD. Making precise predictions from first principles about the behaviour of the infectious agent in humans and animals are not possible. Finding out what it can do can only be ascertained by observation and experiment.

The fundamental difference between an atom bomb going off and BSE exploding into an epidemic in cattle and spreading to cause fatal human infections is that bombs are *designed* using well-understood parameters, whereas BSE is a product of *evolution*. We know nothing about the factors that drove the development of its precursor PrP protein or its change into PrP^{sc} or gave it the ability to cross species boundaries and cause disease in humans. One useful comparison can be made, however. The parameter that explains the *growth* of a nuclear chain reaction is the same as that of an epidemic. In early 1933 the Hungarian scientist Leo Szilard asked Lise Meitner whether he could work with her on nuclear physics. But Hitler took over, so he moved to London. On 12 September as he was crossing Southampton Row the idea occurred to him that if an element split by neutrons emitted *two* when it was split by *one*, then there would be a nuclear chain reaction. Szilard was an inveterate patenter (he filed more than 45 applications with Einstein in the 1920s and 1930s); he thought the idea too dangerous to publish, but patented it in 1934 (UK No 630,726), assigning it to the Admiralty for secrecy. The chain reaction principle also applies to an infection. If a victim infects more than one susceptible contact, it will spread. Less than one, and it will die out. Mathematical modellers call this number R_0. It was greater than one for BSE because material from animals with the disease was fed to many more in MBM, and it is now less than one because MBM is banned, so BSE will eventually disappear.

Following Daubert and MAUD as guide and example will undoubtedly improve science-based policymaking. Filling the centre of the Fleck thought collective with the right kind and number of specialized experts is particularly important, particularly with a problem like BSE where the science is so incomplete and so uncertain. But even when this happens, success is not guaranteed. This is because of the way that a thought collective works. Influence not only travels outwards, from specialized to general expert and then to policy makers, but it goes in the other direction as well. This means that administrative and political considerations are bound to have effects on scientists and on their contributions to policy making. The impact of these effects can be very great.

A dramatic example is the *Challenger* Space Shuttle disaster. The similarities between the events that led up to it and BSE are uncanny. The Shuttle consists of a plane-like orbiter with three engines fed from an enormous fuel tank containing hydrogen and oxygen; 80 per cent of thrust at lift-off comes from two solid fuel rocket boosters (SRB) on either side of the fuel tank. They are built from segments; hot gases are prevented from escaping through the joints between them by two rubber O-rings. At 58.788 s after the launch on 28 January 1986 a small flame emerged from the aft joint of the right SRB. Both O-rings had failed (Figures 11.1 and 11.2). The flame became a plume. The SRB swivelled into the *Challenger*'s right wing, then hit the hydrogen–oxygen tank; the plume was blown by the slipstream onto the tank, which was breached at 64.660 s. It exploded nine seconds later, totally enveloping the *Challenger*, now travelling at Mach 1.92 at

Figure 11.1   Challenger: a few seconds after the explosion.

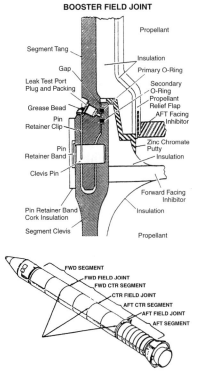

Figure 11.2   Solid rocket booster field joint.

46,000 feet. The crew compartment hit the sea at 200 mph two and a half minutes later. All the astronauts perished.

Bad enough as this very public disaster was, there was surprise, anger, and outrage when the investigatory Commission set up by President Reagan soon found that problems with SRB O-rings were long-standing. Damage to the first ring caused by hot gases—erosion—had occurred on the second shuttle launch, in November 1981, and many times thereafter. Sometimes hot gas had blown past the first ring; in one there had been erosion of the second ring. On the first occasion of hot gases reaching the second ring there had been a three-day cold snap at the Florida launch site, a most unusual event. There was also concern because the second ring, the last defence if the first ring failed, sometimes did not seal because the joint distorted after ignition. In the words of a memorandum sent by Roger Boisjoly, an engineer at Morton Thiokol, the SRB manufacturers, to Robert Lund, its Vice President Engineering, on 31 July 1985,

> it is a jump ball as to the success or failure of the joint because the O-ring cannot respond to the clevis opening rate . . . the result would be a catastrophe of the highest order—loss of human life. . . . It is my honest and very real fear that if we do not take immediate action to dedicate a team to solve the problem with the field joint having the number one priority, then we stand in jeopardy of losing a flight along with all the launch pad facilities.

Boisjoly was a member of an unofficial seal task force set up in July 1985 to solve the O-ring problem. The aim of his memorandum was to get things done faster—in his own words 'to turn the flame up a little bit'.

In October with another SRB engineer, Bob Ebeling, he attended a seminar in San Diego attended by 'probably 600 of this country's finest seal people'. Boisjoly and Ebeling asked 'is there a neat way of fixing this?' But 'not a one came forth'. The task force went on with its work.

The crunch came the day before the disaster. NASA was concerned about the forecast for Cape Canaveral, which predicted unusually cold weather. The Solid Rocket Motor Manager at the Marshall Space Centre in Alabama remembered the previous cold-related O-ring problem. There was a 45-minute telephone conference between Alabama, Cape Canaveral, and Thiokol in Utah. Another followed at 8.15 PM Eastern Standard Time to allow more people to be involved and the faxing of engineering data. Thirty-four engineers and managers from Marshall and Thiokol took part. It lasted three hours. It was a Fleck classic. The engineers were unhappy. Thiokol recommended that launch should not take place unless the O-ring temperatures were 53 °F or more. Marshall managers said that this amounted to a new Launch Commit Criterion. One explained 'My God, Thiokol, when do you want me to launch, next April?' Thiokol withdrew from the conference and went into caucus. Boisjoly said that experience indicated a correlation between low temperatures and O-ring problems. Jerry Mason, the senior manager running the caucus, said that erosion of the O-rings could be tolerated, and that the second rings would operate as they had in pressure tests done below freezing. He said 'we have to make a management decision'. The four senior managers voted. Three went for launch. Robert Lund hesitated. Mason asked him to 'take off his engineering hat and put on his management hat'. He voted for launch. Thiokol rejoined the conference and the launch proceeded, to disaster.

In her incisive analysis, 'The *Challenger* Launch Decision', Diane Vaughan points out that even the O-ring specialist engineers *did not believe the SRB joints would fail*' (her italics). After the conference, for example, Roger Boisjoly called at a colleague's office to ask him to document post-flight damage because the cold launch would give them new data points about low temperature effects. Vaughan's analysis shows that over the years signals of danger were normalized into acceptability. O-ring erosion would continue, but redundancy in the joint design would always save the day. Vaughan identifies belief in redundancy as a scientific paradigm—a synonym for a Fleck 'thought style'. Just like 'remoteness' for BSE, it had a long life and was very resistant to change, despite the accumulation of evidence chipping away at its foundations. As Richard Feynman, former Los Alamos scientist, Nobel Laureate and member of the President's Commission, observed, decision making was 'a kind of Russian roulette ... (The Shuttle) flies (with O-ring erosion) and nothing happens. Then it is suggested, therefore, that the risk is no longer so high for the next flights. We can lower our standards a little bit because we got away with it last time ...'

The 'redundancy' and 'remoteness' paradigms were swept away when the *Challenger* blew up at 16.39 GMT on 28 January 1986 and when Stephen Dorrell made his statement in the House of Commons at 15.31 GMT on 20 March 1996. Their ends were as sudden as these events were dramatic. The Morton Thiokol recommendation faxed 11 h before launch said, among other things, that 'if the primary seal does not seat, the secondary seal will seat'. MAFF officials were preparing to issue advice to the public that 'there is currently no scientific evidence to indicate a link between BSE and CJD' less than three weeks before Dorrell's statement.

Both paradigms played significant roles in accentuating the impact of the disasters. Because launching the Shuttle was perceived to be low risk, space flight was seen as routine. It was now safe to fly non-astronauts like Christa McAuliffe, the teacher on *Challenger*. Her death, on television, gave the disaster a particularly tragic dimension. Because 'remoteness' had driven a policy of over-optimistic reassurance about the safety of beef, the sudden U-turn had a massive impact on belief in the veracity of government and on trust in science.

Chapter 12

vCJD—The Future

What about vCJD? Making predictions about the future size of the epidemic or how long it will last is very difficult because we have no direct information about any of the parameters needed to make the calculation. We do not know how many people were infected before effective control measures were introduced, when they were infected, or the length of the incubation period— the interval between infection and the beginning of the illness. Conservative estimates about the number of infected animals slaughtered for consumption before the November 1989 SRM ban range from 460,000 to 482,000. So exposure could have been massive.

So far all vCJD cases have come from the 40 per cent of the UK population that has a methione codon at position 129 in both copies of the PrP gene, and this fact has to be taken into account by mathematical modellers along with the observation that the mean age of death of patients is 28. The most optimistic predictions about the vCJD epidemic are those of Alain-Jacques Valleron and his colleagues. They assumed that the risk of developing the disease were much greater before the age of 15 and decreased rapidly thereafter (the mean age of death appears to be staying the same as the epidemic progresses, indicating that young people are more susceptible), and that all infections occurred between 1980 and 1989. Their calculations give a mean incubation period of 16.7 years and predict that the total number of cases will be 205. Estimates by a group at the London School of Hygiene and Tropical Medicine used a different approach. They assumed that the hazard of infection was proportional to the incidence of BSE and they looked at a wide range of incubation periods. Modelling using vCJD data reported at the end of 2002 suggests that the number of cases is unlikely to exceed 540; in the next 5 years 40 cases can be expected. Another way to estimate how many people are incubating vCJD is to look for prion protein in tonsils and appendixes removed at routine surgery. It begins to build up in the lymphoid follicles of these organs before vCJD symptoms appear. One positive appendix was found in 8318 samples taken across the UK from patients aged 10-50 between 1995 and 1999. The researchers who conducted this study made no formal predictions because calculations based on a single positive case have enormous statistical uncertainties. As far as they go, estimates from these findings indicate 100 future cases with a possible total of up to 2,600.

Intensive investigations have failed to reveal how people with vCJD became infected. This is not surprising. Finding out what people ate in the few days preceeding a food poisoning outbreak is very difficult; expecting the relatives of those suffering from vCJD to remember accurate dietary details of ten and more years before is hopelessly optimistic. The only statistically significant findings so far provide questions rather than explanations. They are the cluster of five cases in Leicestershire, possibly explained by local butchery practices causing brain contamination of red meat (although it is likely that such practices were widespread in the UK), and an observed excess of cases in the northern half of Britain. People living in Scotland and the North, North West and Yorkshire, and Humberside English regions in 1991 were one and three-quarter times more likely to have developed vCJD than those living in the rest of Britain. A small number of cases of vCJD has also occurred in France and Italy in patients who had never visited the UK; it is presumed that they contracted the disease in their native countries. BSE occurs in both but at a much lower level than in the UK.

vCJD is a new disease. But there is an instructive historical parallel. At the beginning of the twentieth century about 2300 people died every year in Britain from dementia paralytica, a disease for much of its history called 'general paralysis of the insane' (GPI). Variant CJD and GPI have much in common. Both illnesses start with psychiatric symptoms and then, after many months, finish with profound dementia and

WATER-BEDS.

EDMISTON AND SON,

5, CHARING-CROSS (Late 69, STRAND),

LONDON,

Beg to call the particular attention of the Managers of Hospitals and Dispensaries, and the Medical Profession generally, to the price and quality of their Hot or Cold Water-Beds. The prices hitherto charged being so high as to limit the sale to the affluent, they are induced to offer them at such prices as will enable the public generally to realize the advantage and comfort to be derived from their use.

WATER-BEDS,

According to size, £4 4s. to £10 10s.

Figure 12.1   Water bed. Used in the case of terminally ill GPI patients from the mid-nineteenth century onwards.

total paralysis. Charles Mercier, the soft-boned psychiatrist, gave a description of the end stage of GPI in his 1914 textbook. It applies very well to vCJD today. 'The patient lies in bed in the extreme stage of dementia . . . arms crossed . . . legs drawn up . . . incapable of conveying food to the mouth. . . . it is with the greatest difficulty that the patient is saved from getting bed-sores . . . they emaciate, and the emaciation is often extreme' (Figure 12.1). Both conditions are caused by infectious agents and both have very long incubation periods—for GPI ten and more years between infection and the onset of symptoms. Untreated GPI, like vCJD, is always fatal. The behavioural changes at the beginning of either illness present problems for carers. 'Asylum ear' and broken ribs were the nineteenth-century lot of GPI patients. William Wilson had it and in 1869 was kicked to death in the Refractory Ward at Lancaster Asylum by his attendants. Patients with vCJD have also challenged the health care system and found it wanting. They may not have suffered physical violence, but particularly in the early years of the disease their needs were poorly met by hospitals and social services departments. Residential palliative care was sometimes difficult to get because the patients were young and did not have cancer.

During the twentieth century more than 75,000 people in Britain died from GPI. It was caused by a fragile corkscrew-shaped bacterium, *Treponema pallidum*. It causes syphilis. But only a minority of untreated patients with this disease went on to develop GPI. It is not known why they were singled out, or, when they fell ill, why, more often than not, they developed delusions of grandeur. Nothing has changed since the great German psychiatrist Emil Kraepelin (pupil of von Gudden and co-worker with Alzheimer) wrote in 1926,

> the conditions governing the fateful development of the syphilitic affection into general paralysis are unknown.

Old textbooks give lurid examples of what happened. Charles Mercier again:

> In no other form of insanity do we witness such exaggerated hyperbolical exaltation. The patient owns millions and millions...has hundreds of wives, thousands of children....is the greatest inventor, artist, poet, warrior, statesman, pitch-and-toss player the world has ever seen....is lavishly benevolent, giving cheques for millions, written on dirty bits of newspaper, to all bystanders.

The early symptoms of vCJD are very different. Nevertheless, our inability to explain them is as great as it is for GPI.

GPI no longer occurs in Britain. This is not because people have stopped having unprotected promiscuous sex, the best way to catch syphilis. Penicillin killed it off; *Treponema pallidum* is more sensitive to it than any other organism. Prion diseases are very different. They are not caused by bacteria and so antibiotics are useless. The prospects for developing successful treatments are very poor. Lateral thinking and lots of luck will be needed. Again, comparison with GPI is instructive. Julius Wagner-Jauregg was an Austrian psychiatrist. In 1889 he was appointed head of the University Psychiatric Clinic at Graz. He became a great friend of Theodor Escherich, the discoverer of *E. coli*, and called his son after him. In 1893 he moved to a professorship in Vienna. From the late 1880s onwards he began to test the idea that fever might help psychiatric patients. He obtained tuberculin—a product obtained from cultures of tuberculosis bacilli—from Escherich and got interesting results. In June 1917 he tried malaria. Two of the first nine patients with GPI recovered. By 1921 he had treated 200, of whom 50 had been able to return to work. The treatment was first used in Britain in 1923. It prevented death in a third of patients because the fever killed the *Treponema*; half of this group recovered enough to go home. But the other half had been treated too late. Their brains had been irreversibly damaged and they remained in mental hospitals for the rest of their lives. So early treatment was essential. If drugs for vCJD ever became available it will probably be the same.

Figure 12.2   Julius Wagner–Jauregg.

There were other downsides to the Wagner–Jauregg treatment. There is no doubt that it hastened death in more than a few. There was the technical challenge of having a transportable source of malaria readily available. And postcode prescribing was very evident. Whether a Lancastrian or a Yorkist received treatment in the 1920s depended on their mental hospital; Liverpudlians in Rainhill and residents of Sheffield in Wadsley received it but inmates of Lancaster and Stanley Royd did not.

The Wagner–Jauregg treatment was not a complete answer to GPI. But it was the first remedy ever to cure patients permanently of a psychiatric illness. Wagner–Jauregg was given the Nobel Prize in 1927 (Figure 12.2). His achievement was overtaken by Alexander Fleming's penicillin, discovered in September 1928, exactly one year after the announcement of his award. He is now forgotten. Only time will tell whether a similar fate will befall the Nobel committee's judgement about the prion hypothesis!

In his book Ludwik Fleck used syphilis as an example of the close relationship between politics, public opinion, and science. He pointed out that powerful social pressures to control syphilis as a carnal scourge caused governments to spend large sums on research and diagnosis. Listing the famous victims of GPI provides another reason; politicians themselves were affected. Winston Churchill's father, Lord Randolph Churchill, died of it. Not long after his book appeared Fleck received personal confirmation of his ideas in the most horrible way imaginable. He was born, educated, and did his most important work in Lwów. During his lifetime in turn it was in Austria-Hungary (as Lemberg), Poland, the Soviet Union (after the Nazi–Soviet pact), under German occupation, and the Soviet Union again (as L'vov). It is now in the Ukraine (as Lviv). He was trained as a microbiologist by Rudolf Weigl, famous for developing a vaccine against typhus made by the intra-rectal inoculation of lice. By the late 1930s Fleck had already suffered from the anti-semitic policies of the Polish government (his book had to be published in Switzerland) but only when the Germans took over was he put into the ghetto. In its hospital he developed a typhus vaccine made from patients' urine. It saved his life and those of his wife and son. He was deported to Auschwitz in early 1943, to work on typhus in the Block 10 Hygiene Institute. In December 1943 he was transferred to the SS Hygiene Institute at Buchenwald to continue the development of his typhus vaccine (Figure 12.3). After liberation he returned to Poland. It is certain that Fleck's expertise saved his life. The Germans' fear of typhus was greater than their racialism. He was not alone in benefiting from this fear. To make his typhus vaccine Weigl needed human 'feeders'. Every day they strapped about 10 cages containing 400 lice onto their legs. He used Lwów university professors and other intellectuals in a Schindler-like scheme; 'feeders' had passes from the German High Command and got special rations. They often had infected lice on their calves or thighs so the Gestapo gave them a wide

Figure 12.3   Buchenwald. Ruins of the Waffen–SS Fleckfieberserum Institute where Fleck worked on typhus.

berth. In this way Weigl protected the lives of people like Stefan Banach, leader of the Scottish School of Mathematics, named after its Lwów meeting place, Kawarnia Szkocka (the Scottish Café).

Typhus is caused by an organism called *Rickettsia prowazecki*. Like *Treponema pallidum* it is so specialized that it only grows in animal tissues. This makes laboratory diagnosis more complicated than for organisms like *E. coli*, which can be cultured quickly and cheaply on non-living media. However, by chance, infections with both organisms cause patients to develop antibodies which react with easily prepared reagents: tissue extracts in syphilis (the Wasserman reaction) and certain strains of the common bacterium *Proteus* in typhus (the Weil–Felix reaction). Neither test was perfect—not all typhus patients became Weil–Felix reactors and not all Wasserman positives were due to syphilis (smallpox immunization was a well-known cause of 'false positives', hence the traditional advice to applicants for US visas to have the syphilis test before the vaccination) but for many years they were good enough for routine use. The League of Nations took a particular interest in the Wasserman reaction and organized Congresses to perfect it. But, like the Weil–Felix reaction, and the League itself, it was dumped into the dustbin of history a long time ago. Nevertheless, these reactions have relevance to vCJD, where the outstanding diagnostic problem is the lack of a blood test. The prospects are not good. If the prion hypothesis is correct, the only way to diagnose infections with BSE is to look for PrP^{sc}. But all the evidence suggests that it only starts to build up late in the course of the disease, probably not long before the onset of symptoms, and that it is never going to be present in the blood in readily detectable amounts. So diagnosis needs a different approach. One way to do this is by detecting other changes produced by the disease—ones analogous to the Wasserman antibodies in syphilis. The best that has been achieved so far is the measurement of certain proteins in cerebrospinal fluid. One of them, 14-3-3, is helpful in the diagnosis of sporadic CJD. Results with vCJD have been disappointing; its levels have been abnormal in only half the patients tested. So a definitive specific diagnosis of vCJD still rests on the examination of brain tissue. In terms of diagnostic sophistication we are at a point reached by GPI in 1913, when it was first diagnosed by brain biopsy:

> the head was shaved, the front painted with iodine, and a region frozen with ethyl chloride. A revolving dental drill was thrust quickly through the skin and deeper tissues; a few revolutions pierced the skull. After removing the drill a long thin needle was pushed firmly into the cortex of the brain . . . and a small cylinder of grey and white matter sucked into a syringe. Organisms were looked for by dark-field microscopy. . . . in all cases thus far examined, practically no pain has been experienced.

Chapter 13

The Precautionary Principle

BSE presented a very difficult challenge for policy makers. Its novelty was extreme. There had never been a TSE in cattle before. It was not possible to predict whether it could infect other species, including humans. All that could be said at the beginning, using scrapie and CJD as comparators, was that the agent would be very heat resistant and that its incubation period would be very long; finding out rapidly how many animals (and people) had already been infected recently would probably be impossible. Hard scientific information was needed quickly, but BSE had emerged at a bad time. Funding for research on animal diseases had been cut. In any case, a long incubation period meant that getting useful information would be very slow, and that assessing the success or failure of control measures would take a very long time. Normal iterative processes, even ones operating on annual cycles of review, would not work well. Taken together, it could be said that all these things made BSE an ideal case for the application of the precautionary principle.

The principle is commonly held to have emerged during the development of European environmental policies in the 1970s, although John Snow's removal of the Broad Street pump handle in 1854 has been quoted as its first application. It was enshrined at the 1992 Rio Conference on the Environment and Development as principle 15 of the Rio Declaration: 'In order to protect the environment, the precautionary approach shall be widely applied by States according to their capability. Where there are threats of serious or irreversible damage, lack of full scientific certainty shall not be used as a reason for postponing cost-effective measures to prevent environmental degradation'. Its application is not restricted to environmental issues; in its 2000 'Communication' on the principle the European Commission indicates its appropriateness for the protection of human, animal and plant health as well.

The first European legal decision that used it concerned BSE. In 1998 the European Court of Justice upheld a Commission ban on the export of British beef. It said that because of 'the seriousness of the risk and the urgency of the situation, the Commission did not react in a manifestly inappropriate manner by imposing, on a temporary basis and pending the production of more detailed scientific information, a general ban on exports of bovine (products)'. This was not the first time that the principle had been used to protect against BSE. Sir Richard Southwood's recommendation that clinically ill animals should not go into the food chain and John MacGregor's decision to extend the Southwood offal ban from babies to everyone went further than the science. These were good decisions. But the perception that they were precautionary when judged against 'remoteness' had a negative effect. Without doubt, it contributed to the lack of rigour in the implementation of control policies in general; the lack did not matter because there was perceived to be a substantial overkill in delivering safety.

But for novel pathogens the raw precautionary principle had already shown itself to be a bad guide for policy makers. David Lewis was an 18-year-old new recruit to the US Army at Fort Dix in New Jersey. On 4 February 1976, like many of his fellow recruits during the previous three weeks, he had developed a respiratory infection. The medical officer told him to go to bed. He went instead on a five-mile forced night march, collapsed, and in spite of emergency measures including mouth-to-mouth respiration by his sergeant, died before the night was out. He had been killed by influenza. Within a week the virus had been identified as type A/swine. Three other soldiers had also been infected by it, with much milder consequences; the other cases at Fort Dix had another much commoner human virus, A/Victoria. A lethal swine virus sent shivers down the spines of the virologists. There was a real possibility that Lewis's virus was closely related to the one that had swept the world in 1918 killing at least 20 million people—antibodies from the blood of survivors had shown the swine connection in studies some years before. The Center for Disease Control (CDC) at Atlanta mobilized. Virological results were confirmed on 13 February. Vaccination was discussed at a meeting on the 14th. Work started on a recombinant vaccine on the 17th. There was a press conference on the 19th. The media made links between Fort Dix and 1918 on the 20th. By 13 March, David Sencer, CDC's Director, had prepared a memorandum entitled 'Swine Influenza—ACTION'. It went up the line to Denis Matthews, Secretary of Health, Education and Welfare (HEW) and on to President Ford. He was briefed on 15 March. On 24 March a meeting of a 'Blue Ribbon Panel' of experts, including the polio vaccine Nobel Laureates Jonas Salk and Albert Sabin, was convened at the White House. Sencer gave a briefing. The President turned to Salk, who strongly urged mass immunization. Sabin followed, and agreed. Ford was impressed. He knew that Salk and Sabin hated each other. There was a show of hands. All went up. To be certain, the President suspended the meeting and went into the Oval Office to allow any doubters to discuss their worries with him in private. There were none. The whole process was a Fleck classic. The White House swine flu 'thought collective' had an inner circle of esoteric experts—the flu virologists, an outer circle—Salk, Sabin, and Sencer, and exoterics, like officials from HEW. Sencer's memorandum linked them all. It was based on what solid science there was, but it oversimplified. It was vivid. Its main pessimistic assumption had the flavour of certainty: 'present evidence and past experience indicate a strong possibility that this country will experience widespread A/swine influenza in 1976-77'. It put a gun to President Ford's head by saying 'the situation is one of 'go or no go'. . . . A decision must be made now. . . . A public health undertaking of this magnitude cannot succeed without Federal leadership. . . . ' Ford was in the middle of difficult primary elections for the Republican Presidential nomination. Ronald Reagan was doing well. Ford needed to shake off his reputation as a bumbler. He went on TV in time for the evening news. Flanked by Sabin and Salk he said 'I am asking Congress to appropriate $135 million, prior to their April recess, for the production of sufficient vaccine to inoculate every man, woman and child in the United States'. Work started at once. So did trouble. Who would indemnify if things went wrong? The vaccine manufacturers were very unhappy. So was Congress. A Bill ensuring Federal funding was passed—eventually—but only after Legionnaires disease had struck in August in Philadelphia and sharpened minds. The first vaccine shots went in on 1 October (Figure 13.1). On the 11th three elderly Pittsburgers

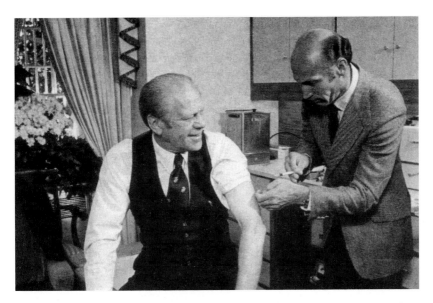

Figure 13.1   14 October 1976. President Ford demonstrating the safety of the swine flu vaccine on television.

died shortly after being inoculated at the same clinic. The body count began. On 12 November a case of the Guillain–Barré syndrome was reported in a Minnesota vaccinee. This is a severe paralytic illness known to occur as a rare, delayed, complication of infections. More Guillain–Barré cases came to light at the beginning of December. It was looking reasonably certain that the vaccine was implicated. The programme was suspended on 16 December after 40 million Americans had been vaccinated. It never restarted.

Swine flu never spread from Fort Dix, and the vaccination programme was unnecessary. The media and the public perceived it to be a debacle. It damaged the reputation of the US public health system. Within three weeks of Jimmy Carter becoming President, Sencer was dismissed. Not long after taking office, Joseph Califano, the new Secretary of HEW, asked the Harvard policy analysts Richard E. Neustadt and Harvey Fineberg to report on the programme and identify the lessons that could be learned. He saw two big problems: 'How shall top lay officials, who are not themselves expert, deal with fundamental policy questions that are based, in part, on highly technical and complex expert knowledge—especially when . . . 'the facts' are so uncertain', and 'how should policymakers . . . seek to involve and to educate the public and relevant parties on such . . . issues'. Neustadt and Fineberg identified five 'critical phenomena'. They enthusiastically endorsed the major recommendation of US General Accounting Office in its own report on swine flu that 'when decisions must be made on very limited scientific date, HEW should establish key points at which the programme should be formally *re-evaluated*'. They concluded that '*implementation* is not only something to be done after decision, it is as much or more a thing to think about before decision'. They said that not enough thought had been given to the *media*; anticipating coverage was badly done. They criticised CDC, and Sencer in particular, for mortgaging its *reputation*. The scientific issues had not been posed candidly,. Sencer had been too much of a salesman for his arguments rather that a 'a technician (openly) serving up to his superiors the data for their judgement'. And finally they emphasised the 'slippery' nature of influenza—its changing character, and the enormous degree of ignorance that existed about its natural history. What a basis, they said,

> on which to expose 40 million people to an unknown risk of side effects. And all this on the word of experts, *overconfident* in theories . . . without any proper review of their logic by disinterested scientists. It is not that conclusions were inconsistent with evidence, but that the paucity of evidence belied the force with which conclusions were advanced.

They pointed out that the fundamental differences between polio and influenza were very great and using polio vaccine (Salk and Sabin's expertise) as a model for influenza was not only wrong, but dangerous.

The relevance of the five 'phenomena' to BSE is uncanny. It is a great pity that Neustadt and Fineberg's 1978 government report *The Swine Flu Affair: Decision-Making on a Slippery Disease* or its popular expanded and annotated paperback version had not been obligatory reading for UK health and agriculture ministers and officials in the 1980s and early 1990s. If it had been maybe Southwood would have been *reviewed*, and maybe policymaking would not have been left so often to SEAC, which did not have the background knowledge to assess the *implementability* of some of the things it was recommending, and the scrutiny of policy implementation would have had a high priority and been better done. Cordelia Gummer might not have been asked to eat a superheated burger as an ill-judged *media* stunt, the *reputations* of experts, politicians, ministries, advisers and farmers might stand higher, and the optimistic assumptions coming from the *overconfident* adoption of the scrapie hypothesis might have been challenged earlier.

Since the 1970s much has been written on the precautionary principle. The European Commission's prepared a 'Communication' on it in 2000. It highlights two of Neustadt and Fineberg's major recommendations; the need for measures to be periodically reviewed in the light of scientific progress, and the need for them to be based on an examination of potential benefits, costs and efficiency. These are guidelines for applying the principle; when to use it is the difficult decision. The 'Communication' says that the principle 'is particularly relevant to the management of risk . . . judging what is an "acceptable" level of risk for society is an eminently *political* responsibility'.

Microbial risks have been political for as long as politics itself. On 24 April 1497 the Court of the Baillies of Aberdeen met and tried to do something about prostitutes and syphilis. The Council Register records that on

> The said day it was statut and ordainit be the Alderman and Consale for the eschewin of the infirmites comin out of France and strange parts that all licht weman be chargit and ordainit to desist fra thair vice and syne of venerie and all their buthes [shops] and houses skalit [closed] and thai to pass to wark for thar sustentatione under the payne of ane brand of the yrne (iron) on ther cheks and bannysone [banishment] of the toune.

Half a millennium on, the 'infirmity out of France' has been controlled by penicillin, but the 'licht weman' are still busy in Aberdeen's red light district and the 'Consale' still spends time regulating their activities. To their credit, over the last 500 years the councillors have learned that legislating against sex is unprofitable and their concerns now extend to the risks run by sex workers as well as by their customers. On the national scene things are more dramatic. Microbes exert their most powerful political influence through the Lardner effect. Sudden multiple well-publicized deaths in an outbreak stimulates a public call for action. Scapegoats must be sought. Lessons learned. There must be a full public inquiry.

Chapter 14

BSE, vCJD, and *E. coli:* The Aftermath

The inquiries that follow disasters in England are very English in their administrative heterogeneity. The legal structures that underpin them are labyrinthine. They proceed in many different ways, some in public, some in private, some under oath with advocates and some without lawyers at all. Traditionally, scandals and disasters were investigated by Parliamentary Committees. But party politics intervened often enough to discredit their impartiality, particularly when investigating the conduct of the government. Their coup-de-grâce came in 1912 when the Committee inquiring into allegations that members of the government had corruptly favoured the Marconi Company divided on party lines. The Tribunals of Inquiry (Evidence) Act 1921 was born in consequence. Inquiries under the Act are set up by resolutions of both Houses of Parliament to investigate matters of 'urgent public importance'. Their powers are essentially those of the High Court. No doubt they are what many have in mind when they call for 'a full public inquiry'. But, like other 'grand inquiries' with strong legal powers made possible by laws like the Regulation of Railways Act 1871 (resulting from the Armagh disaster), very few—less than 20—have been held. The vast majority of public inquiries have been set up by government ministers; many have not been statutory and so have not had the power to compel witnesses to attend. The BSE Inquiry fell into this category. The composition of its committee followed the traditional English pattern of health disaster inquiries. Like Croydon (Typhoid, 1937), Devonport (Contaminated Infusion Fluids, 1972), London (Smallpox, 1973), and Stanley Royd (Food Poisoning, 1984) it was chaired by a senior lawyer assisted by two experts (for BSE Lord Phillips, Master of the Rolls, Ms. June Bridgeman, a retired civil servant and Professor Malcolm Ferguson-Smith, a medical geneticist). It differed from its predecessors in its massive breadth and its duration (it started work in January 1998 and its 16-volume report went to the printers in October 2000). Its remit was

> To establish and review the history of the emergency and identification of BSE and new variant CJD in the United Kingdom, and of the action taken in response to it up to 20 March 1996; to reach conclusions on the adequacy of that response, taking into account the state of knowledge at the time; and to report on these matters to the Minister of Agriculture, Fisheries and Food, the Secretary of State for Health and the Secretaries of State for Scotland, Wales and Northern Ireland.

It was unusual in being restricted in its forward look, and having its proceedings time limited; the inquiry was to report by 31 December 1998.

The scale of its trawl through government files was unprecedented. Transcripts of hearings went onto its website within hours; during the lifetime of the inquiry it received more than 1.5 million page requests.

Phillips concluded that

> because the right policy decisions were taken, BSE is today within reach of eradication and millions have received a high degree of protection from the risk of ingestion of potentially infective products or by-products of the cow…plaudits must, however, be muted. Not all went well. All too often the correct policy decision was marred by:

- the time that had been taken to reach it;
- the lack of rigour in considering how to give effect to it;
- lack of rigour in implementing, enforcing and monitoring the Regulations introduced to give effect to it.

The report listed the lessons learned in 132 bullet points. The largest number addressing one topic—32—concerned the use of scientific advisory committees. Committees should include experts in the area; advice should be in terms understandable by a layperson and it should normally be made public and state the reasons for its conclusions; and government departments should have the expertise to understand and evaluate the advice were the important ones. The report highlighted the malignant effect induced by the perception that precautionary measures went over-the-top. Four bullet points drew an identical lesson: 'when a precautionary measure is introduced, rigorous thought must be given to every aspect of its operation with a view to ensuring that it is watertight' (lessons from the introduction of the ruminant feed ban); 'when a precautionary measure is introduced, rigorous thought must be given to every aspect of its operation with a view to ensuring that it is fully effective' (lessons from the introduction of the animal SBO ban); 'when a precautionary measure is introduced, rigorous thought must be given to every aspect of its operation with a view to ensuring that it is fully effective and its purpose and application understood by those concerned' (lessons from the introduction of the human SBO ban); and 'precautionary measures should be strictly enforced even if the risk that they address appears to be remote' (lessons from the emergence of vCJD).

Readers of an inquiry report turn first to the chapter that criticizes named individuals. Phillip's opprobrium fell most often on members of the outer ring of the esoteric circle of the government's BSE thought collective; twelve government vets, eleven government medical officers, and the Southwood committee. Eleven senior civil servants and five government ministers in its exoteric circle were individually blamed. That over half of those named in the report were doctors or vets is right. Their professional backgrounds gave them the knowledge and standing to challenge the thought collectives' optimistic position and the pillars which supported it, 'remoteness' and the scrapie hypothesis. Not only did they fail to do this, they participated fully in a policy of 'not rocking the boat', 'sedating' the public, and 'leaning into the wind'—a process described by Brian Dickinson, Head of the MAFF Food Safety group, as making 'more reassuring sounding statements than might ideally have been said'. Even if they were constrained because their membership of the esoteric circle carried with it, as Fleck said, a requirement to vigorously defend its thought style—in civil service jargon, the 'line to take'—they should have known better. The doctors in particular were clever, thoughtful and independent-minded. The English Chief Medical Officers Donald Acheson and Kenneth Calman were public servants of the highest quality and integrity. They genuinely believed that their actions were in the public interest. Their 'leaning into the wind' entrapment shows the power of a well-established thought-style. It would have been better if they had been familiar with Leo Szilard's story about a conversation between the Nobel prize-winning physicists Enrico Fermi and Isidore Rabi held soon after Lise Meitner and Otto Frischs' discovery of nuclear fission:

> Fermi said, 'Well ... there is a remote possibility that neutrons may be emitted in the fission of uranium and then of course perhaps a chain reaction can be made.' Rabi said, 'What do you mean by "remote possibility"?' and Fermi said, 'well, ten per cent.' Rabi said, 'Ten per cent is not a remote possibility if it means that we may die of it. If I have pneumonia and the doctor tells me that there is a remote possibility that I might die, and its ten per cent, I get excited about it.'

The Phillips Inquiry cost more than £30m. It went on a year and nine months longer than originally planned. Other long-running multimillion pound inquiries were set up at about the same time—Learning from Bristol (children's heart surgery at Bristol Royal Infirmary) established in June 1998 which cost £14m, and the Bloody Sunday Inquiry established in January 1998, whose costs are currently estimated at £120m. All had remits to inquire into events long since past. A very different process was followed with *E. coli* O157 in Wishaw. Two inquiries were set up while the outbreak was still in progress. My expert group met in private. Our remit was 'to examine the circumstances which led to the outbreak in the central belt of Scotland and to advise on the implications for food safety and the general lessons to be learned'. The Secretary of State for Scotland, Michael Forsyth, asked for priority recommendation to be made to him by the end of 1996—33 days after our establishment on 28 November. On 5 December the Fatal Accident Inquiry was announced. There was a Scottish food poisoning precedent for these arrangements. During the second week of August 1922, six guests and two boatmen at the Loch Maree Hotel in the Scottish Highlands died of botulism. They had all eaten potted wild duck paste sandwiches at lunch on the 14th. Loch Maree is very remote; but the toxin-containing duck paste was made by a big food-processing company in England. As after Wishaw, an expert report was prepared for the Scottish Office, and a Fatal Accident Inquiry was held. There, however, the resemblance ended. The politics were different. In August 1922 Lloyd George was still a confident Prime Minister. But only eight months of 1996 had separated Wishaw from Stephen Dorrell's announcement in the House of Commons about the link between BSE and vCJD. A general election was imminent. John Major's government was in serious difficulties. Not only was Michael Forsyth's own seat at Stirling under threat, but there was a real prospect of electoral melt-down for the Conservative Party in Scotland. These pressures had modulated Forsyth's doctrinaire right-wing views into pragmatism. His unionism was now wrapped in tartan of the loudest hue. There is no doubt that these forces were key in his decision to set up my group. The speed and vigour of his response—which had all party support—could be seen as distancing the Scottish Office from the BSE débâcle and the government departments involved, Health and Agriculture south of the border. It would do no harm to his reputation as an effective minister. Our 'lessons learned' remit would apply to the UK as a whole and so we were challenging the traditional lead roles of the London departments. They certainly saw it that way. My group debated the handling of this issue. Eventually we decided that it would be wise to invite observers from the Department of Health and MAFF to attend our meetings. As expected, they resisted our regulatory proposals with vigour. But to paraphrase an old saying, it was better to have them inside the tent, rather than outside pissing on to us from a great height. We met with the Advisory Committee for the Microbial Safety of Food, which has a UK remit, in London at the Department of Health. They also showed their displeasure to us as usurpers.

My group made many recommendations about the prevention of *E. coli* O157 infections. They ran from farms, through slaughterhouses, meat production premises and butchers shops to the point of consumption; enforcement, surveillance, research, and the handling and control of outbreaks were also covered. There was nothing new or revolutionary about them. Many were just plain common sense. There was a central theme, the accelerated implementation of HACCP. Our report devoted a chapter to it. HACCP stands for Hazard Analysis Critical Control Points. We described it as

> a structured approach to analysing the potentional hazards in an operation; identifying the points in the operation where the hazards may occur; and deciding which points are critical to control to ensure consumer safety. These critical control points (CCPs) are then monitored and remedial action, specified in advance, is taken if conditions at any point are not within safe limits. HACCP is both a philosophy and a practical approach to food safety.

HACCP first emerged as a system to provide safe food for astronauts. The prospect of diarrhoea in zero gravity was too horrible to contemplate. NASA joined with the Pillsbury Company and the US Army to develop it. Since the early 1970s it has been adopted worldwide, underpinning food safety legislation in the US and in Europe. Being in favour of it is uncontentious. Its approach is utterly rational and its underlying philosophy of self-regulation fits the modern trend. Hardly anyone disagrees with the proposition that an inspector's time is better spent in making sure that a food business has an effective system for monitoring hazards, does something about them, and can prove it by good record-keeping, rather than by the examination of walls and ceilings as surrogate measures of safety. In addition, we were in very good company. Lord Cullen, in his Piper Alpha Inquiry Report, recommended a philosophically identical way forward for offshore installations. The main difference is that it is called a 'Safety Case' rather than a 'HACCP'. He said that

> operators...should be required by regulation to carry out a formal safety assessment of major hazards, the purpose of which would be to demonstrate that the potential major hazards of the installation and the risks to personnel thereon have been identified and appropriate controls provided....It is a legitimate expectation of the workforce and the public that operators should be required to demonstrate this to the regulatory body.

Everybody agrees that HACCP is a good thing. So why was it necessary for my group to recommend that HACCP implementation in the UK should be speeded up? We were fighting a government attitude that was the same as St Augustine's to celibacy: 'Give me chastity and continency, but not yet.' There was a palpable lack of enthusiasm about HACCP. Official expertise seemed to reside in enumerating implementation difficulties rather than finding solutions to them.

We knew that the implementation of HACCP in a small food business was not easy. For its proprietor to identify hazards he had to have at least a basic knowledge of food safety. People like Mr Barr clearly did not. We agreed with the opinion that he was not alone. Many butchers were like him. But we did not accept that the right response to this problem was to move slowly. Traditional inspection had failed in Wishaw. People had died. So as an interim measure to help both butchers and inspectors we proposed the licensing of butchers like Barrs—ones handling both raw meat and unwrapped ready to eat foods. To get a licence from the local authority a butcher would have to meet various conditions along HACCP lines. If the prospect of doing this caused the butcher to develop his own HACCP, so much the better. Licensing was not a new idea. The powers to introduce it were in the 1990 Food Safety Act. But it had been totally incompatible with the deregulatory ethos of Conservative governments—until Wishaw.

We delivered our interim report a few hours before expiry of the deadline, Hogmanay 1996. Gaining access to the Scottish Office at St Andrew's House in Edinburgh was difficult because the Labour-controlled city council had caused a fun fair to be set up outside its front door. Forsyth reported our recommendations to Parliament in mid-January, embedding them in the government machinery, and our final report went through the same process in April, just before the general election. Our inquiry cost £45,000. Although intensely political, it was never party political. Neither Michael Forsyth or the Conservatives benefited from it—Scotland became a Tory-free zone after the election. The Fatal Accident Inquiry opened its doors on 20 April 1998 at the Gospel Literature Outreach Centre in Motherwell. The manager provided waiting witnesses with reasonably priced food in its restaurant accompanied with gentle unassertive evangelical propaganda. It was a particularly appropriate environment for seeking truths under oath, and for quiet but concentrated catharsis. Evidence was led and submissions heard over 39 days; it concluded its business on 25 June and the Sheriff published his determination on 14 August.

Were the BSE Inquiry, the Pennington Group, and the FAI worth it? Louis Blom-Cooper, a lawyer who has conducted 11 inquiries, listed five reasons for setting them up in his 1992 report on Ashworth Hospital. Mapping achievements against them answers the question.

First, 'horror or disquiet need to be assuaged quickly. By announcing an immediate inquiry, the Government recognises the importance of the matter and assures the public of speedy investigation.' Together, the Wishaw inquiries satisfied these requirements, despite early calls for a full public inquiry. Forsyth's speed also met another need, in that it kicked an acute problem into touch. Public disquiet about food safety could be answered by saying 'wait for Pennington'. But for BSE the speedy establishment of an inquiry to do this was never an option. The crisis developed slowly. The rate of its development was proportional to its incubation period. By the time cases of vCJD appeared, concerns had been aired and debated for so long that the public had been immunized again and again to respond rapidly and vigorously to developments, whatever their kind or importance. The best analogy is the Three Mile Island nuclear reactor accident in 1979. The US Atomic Energy Act of 1946 said

> that the development and utilisation of atomic energy shall be directed towards improving the public welfare, increasing the standard of living, strengthening free competition among private enterprises so far as practicable, and cementing world peace.

Hardly anyone disagreed. The first commercial reactor went on line in 1957. Until the late 1960s antinuclear protest focused on bombs. But at the beginning of the 1970s local anti-power station groups in the US began to proliferate. Their growth matched that of the planned expansion of nuclear power. There was the Environmental Coalition on Nuclear Power in Pennsylvania (1970), Ralph Nader's Critical Mass (1974), and later in the 1970s the Clamshell Alliance in New England, the Abalone Alliance in California, the Crabshell Alliance in Washington, and the Catfish Alliance in the southern states.

On 28 March 1979 a series of failures in the cooling water systems of Unit 2 at Three Mile Island coupled with operator errors caused massive damage to the reactor. Nobody died or suffered physical harm. There was no China syndrome. The release of radioactivity was trivial. But the subsequent report to the US Nuclear Regulatory Commission concluded that 'like...the Watergate Complex (and) the Texas Schoolbook Depository in Dallas the "towers" at "TMI" have slipped into an un-projected half-life as reminders of steep depressions in our national lifeline. Tourists drive by slowly....Three Mile Island is a big deal; something important happened here'. During the incident 165,000 people from Lebanon, Dauphin, York, Lancaster, and Cumberland Counties evacuated the area. The contrast with the world's first big reactor accident could not be greater. The fire in Windscale Pile No. 1 in early October 1957 at Sellafield in northern England released 1250 times more radioactive iodine into the atmosphere than TMI (Figures 14.1 and 14.2). Neither it nor its sister 1940s vintage reactor with their 410 ft high iconic chimney stacks ever worked again. But public reaction was muted. Nobody evacuated Cumberland, England, during the incident. The strongest criticism came from the specialist weekly journal *The Economist* that there would not be a public judicial inquiry. The publication of a summary of an internal inquiry was enough to satisfy.

Blom-Cooper's second and third reasons for holding an inquiry can be taken together. 'Frequently, the events to be investigated involve allegations of fault by government or public authorities, so it is important to allay fears of a "whitewash", by providing for public inquiry under an independent chairman', and 'disasters and events causing public anxiety go beyond the interests of individual victims who may have a cause of action; they require deeper and more searching inquiry'. The Phillips inquiry satisfied these effortlessly; my own inquiry team was criticized for including Stephen Rooke, a Scottish Office official. At the end of the day our report was not found wanting on this account.

Blom-Cooper's fourth reason is that 'a public inquiry gives an opportunity to all who reasonably have an interest in making representation to do so. It thus has a cathartic effect for victims' relatives and, via the media, the public in regard to distress, recrimination, speculation and rumours.' The FAI and Phillips served these functions well. Relatives, and lawyers acting for them gave evidence to both. Phillips also had the difficult task of passing judgement on alternative theories explaining the origins and causes of BSE. The idea that organophosphates—compounds belonging to the nerve gas family—were responsible was put forward with vigour by Mark Purdey, a farmer. Politicians supported him. But the evidence did not.

Figure 14.1   Three Mile Island. The nuclear reactors are located in the circular containment buildings. A cooling tower is at the rear.

'A committee ... as well as establishing the facts and possibly assigning responsibility, can make recommendations to avoid recurrence', was Blom-Cooper's final reason for holding an inquiry. Making recommendations is easy. Getting them accepted is less so; ensuring implementation can be very difficult. Political circumstances drove acceptance of my group's recommendations. We knew that our relatively radical approach would find favour with the Scottish Office, because its observer on the group told us that it would. Implementation of licensing was a different matter. It did not come into force in Scotland until October 2000. To get a licence butchers could follow one of the two approaches outlined in our recommendations; in England they had to have a HACCP. These national differences give a clue to why implementation took so long. Much time was spent by officials trying to reconcile big disagreements north and south of the border. But in historical terms a three and a half year

Figure 14.2   The Windscale Piles (nuclear reactors), No 1 on the left. One ton of air was blown through each reactor and up the chimney every second. Buildings designed by J. Cecil Clavering, the perfector of the 'Odeon' style of architecture. For George Thomson see p. 176.

interval between recommendation and implementation was progress of outrageous rapidity. Consider football grounds. Their licensing was first suggested in 1924 by the *Report of the Departmental Committee on Crowds* set up in response to organizational deficiencies and injuries sustained by many of the 127,047 attendees at the first Wembley Cup Final. Nothing was done. In 1946, 33 people were crushed to death and more than 400 injured at another FA Cup match, this time at Bolton. The subsequent Inquiry recommended local authority licensing. The government did not accept it. Its need became apparent again in 1971 when 66 died at Ibrox, Glasgow. The 1972 Inquiry recommended licensing; it was only applied to top clubs. Extension to all did not come until the Bradford City Stadium fire in 1985, which killed 56.

Only time will tell whether butchers' licensing and the accelerated implementation of HACCP will bring their expected benefits in full. Likewise, it will be a long time before one can be absolutely certain that BSE has ceased to threaten human health. Nevertheless, it is reasonable to suppose that improvements in food safety are going in the right direction and at good speed.

Dying from *E. coli* O157 and vCJD is particularly horrible. Sufferers were not, however, the only casualties. The circumstances surrounding their deaths, particularly the long prelude of false reassurances about BSE, caused enormous damage to the reputations of the government, experts, and scientists as dispassionate guardians of the public interest. Trust was destroyed. Onora O'Neill started her 2002 BBC Reith Lectures 'A Question of Trust' by quoting Confucius: 'three things are needed for government: weapons, food, and trust. If a ruler can't hold on to all three, he should give up the weapons first and the food next. Trust should be guarded to the end: without trust we cannot stand'. She said that Confucius' thought still convinces. Regaining trust was the main reason for the establishment of the Food Standards Agency. Philip James, Director of the Rowett Research Institute in Aberdeen, was asked by Tony Blair in March 1997 to make recommendations on its functions and structure. They were incorporated with little change into a White Paper. The Bill establishing the Agency received the Royal Assent in November 1999 and it started work in April 2000. Management of food safety and standards was removed from MAFF, the Department of Health and their devolved equivalents. The Agency is run by a Board which has its formal meetings in public; it has the right to publish the advice it gives to ministers. Is it regaining trust? When asked 'was the French Revolution a success?' Chou-en-lai replied 'it is too early to say'. This is the right response. It is certain, however, that one will not have to wait so long for an answer. The Agency works much more openly than its predecessors. It has turned its face against 'sedative' policies. It audits inspectors. These are all good things. But its real test will come with the next big food scare.

The Phillips report focused on the working of the government machine. It found many deficiencies. Have all the important lessons been learned? Not yet. Take foot and mouth disease (FMD) in the UK in 2001. The outbreak was one of the biggest ever recorded. But the virus should be much simpler to control than BSE. It was discovered in 1897. There is no controversy about its natural history or its basic properties, which are better understood than for most other animal viruses. It is easy to kill with heat and disinfectants. It has a short incubation period—from two days to two weeks—so the effectiveness of control measures can be judged quickly. The laboratory diagnosis of infection is quick, accurate, and sensitive. Government control policies go back to the second half of the nineteenth century and an excellent state-funded laboratory to work on it was established in Surrey in 1924 at Pirbright. But the deficiencies that emerged during the outbreak were uncannily similar to those exposed by BSE. The FMD contingency plan was constructed to meet European Directive 90/423/EC—to cope with a 10-case outbreak. It was hopelessly optimistic and totally inadequate. Fifty-seven premises in Scotland, Wales, and northern, southern, western, and eastern England had been infected by the time the first diagnosis had been confirmed on 20 February 2001 and by the end of the outbreak in September 2026 premises had been declared infected. As with BSE, the Welsh had already challenged the adequacy of the plans. The Chief Government Vet there concluded a July 2000 memo by telling London that he would continue a Welsh training programme 'because if a problem should arise here in Wales it will be my head that is on the block—in view of the BSE problem not a position I care to contemplate too closely'. At the heart of the problem in controlling the virus, as with BSE, was the exclusion of specialist scientists from the policy-making process. The contingency plan had been prepared without involving the Pirbright laboratory, the World Reference Centre. Scientific inputs about mathematical models and vaccines should have fed into its construction. Because they were not, debates about them raged during the outbreak itself—the worst possible time. Once again, the centre of the esoteric circle had been empty at a crucial time, not because of the lack of first-class British scientists, but because policy makers had still not learned the importance of using high quality scientific information to the full. The contingency plan even failed to mention the sources of scientific advice that could be tapped in an outbreak.

Using the precautionary principle was for BSE neither straightforward or universally beneficial. It was the same for FMD. Rural footpath closures proceeded with almost universal acclaim very early in the outbreak. In a debate on it in the House of Commons on 28 February the Minister of Agriculture suggested that local authorities should 'act on a precautionary basis'. They were given powers of enforcement. But shutting the countryside had not been done on the basis of any scientific assessment of risk, and the process of reopening had not been thought through. But the economic impact of closure on tourism—which generated more rural income than farming—was very great. Farmers whose animals were compulsorily slaughtered got compensation. Bed and breakfast operators, the Youth Hostel Association, shopkeepers, and other rural businesses did not. Compensation and other payments to farmers were estimated at £1.4 billion. Tourism and supporting industries lost revenues of between £4.5 and £5.4 billion. Blanket footpath closures were a mistake. There is no evidence that they prevented virus spread, but plenty that they caused economic damage.

Its handling of FMD caused the competence of the government to be questioned. There were many calls for a public inquiry. It is ironic that when the government successfully argued against having one, before Lord Justice Simon Brown and Mr Justice Scott Baker, it used what it considered to be negative aspects of Phillips in justification. Some farmers, vets, and hoteliers, supported by a substantial part of the national media industry, had taken it to court. In defence of its decision government witnesses like Brian Bender, Permanent Secretary to the Department of Environment, Food and Rural Affairs (previously MAFF) said that there was a wish to learn lessons as rapidly as possible and that a closed inquiry process would lead to greater candour from witnesses and would save financial and human resources. There is some merit in these arguments. But such inquiries run a big risk of failing the test set by Mr Justice Sheen in his report on the capsize of the *Herald of Free Enterprise*: 'In every formal investigation it is of great importance that members of the public should feel confident that a searching investigation has been held, that nothing has been swept under the carpet and that no punches have been pulled.'

The cat was let out of the bag by Ann Margaret Waters, Head of the BSE Liaison Unit, when she said in her statement to the court 'the pressure on witnesses who gave evidence in oral hearings...was immense...intensified by the fact that the oral hearings were in public with a constant media presence and in front of an audience that frequently contained members of the families who had lost relatives to vCJD...' The public nature of the (Phillips) Inquiry did mean that witnesses did not offer views and express opinions as much as they might have done had there not been such public scrutiny of what they said. She said that lawyers advising witnesses encouraged defensiveness.

Clearly, whatever good has come out of Phillips, it has immunized the civil service against public inquiries. It could be said that the big lesson it has learned is to work hard to prevent such public exposure of its activities ever happening again. It is tempting to speculate, although there is no evidence one way or the other, that it may have played a part in influencing the Secretary of State for Health, Alan Milburn, to announce on 12 April 2000 that the inquiry into the mass murderer Dr Harold Shipman would be held in private because it would be quicker and would yield a far greater depth of information by being shorn of 'adversarial features and the distractions of media interest'. There was a judicial review. Lord Justice Kennedy and Mr Justice Jackson were not impressed with these arguments. After taking the circumstances of Shipman's horrendous doings into account, in an unprecedented judgement they said that the decision was irrational. It was set aside in favour of a public inquiry.

BSE and *E. coli* O157 are seen by many as examples of the threats to human existence caused by our modern way of life. The book *Risk Society* by the German sociologist Ulrich Beck is the classic exposition of this view. He claims that the risks posed by late modernity are novel. They are no longer local. They are invisible, may cause irreversible damage, and can only be understood in terms of the scientific knowledge about them. They ignore national boundaries and affect the rich as well as the poor. Their effects are potentially catastrophic. His analysis focuses on radioactivity and toxins and pollutants in the air, water, and food. Beck also claims that today's risks are different from those of the past because they are products of industrialization and *over*production, rather an *under*supply of hygienic technology, and because they are global. He says that 'the movement set in motion by the risk society . . . is expressed in the statement: *I am afraid*!' Popular belief in Beck's story is very strong. People are indeed fearful. The voices of strong proponents like Friends of the Earth and Greenpeace are heeded with respect by many. The popularity of organic food owes a lot to a desire to return to a pre-modern golden age of happy peasants, pathogen-free manure, and an ever-beneficient nature. For Germany Beck said that the transformation to the risk society had achieved real momentum by the 1970s. The same chronology applies to Britain. This fits the timeline of development of *E. coli* O157 and BSE very well. BSE is seen as a brilliant example of the post-modern hazard. Opponents of genetically modified foods use it as an example of science going too far. Those unhappy about farmers being subject to market forces identify MBM-mediated cannibalism as a consequence of late twentieth-century capitalism. Forces favourable to industry are blamed for deregulating rendering, thus allowing the infection to spread. But like Beck's own analyses these opinions are either profoundly ahistorical or grossly oversimple. Some are just myths; it is difficult, for example, to deregulate something, like rendering, that was never effectively regulated in the first place.

Science had nothing to do with the introduction of artificial cattle feeds containing cattle material, which goes back to the beginning of the twentieth century and before, originally to tide animals over the winter. Then farmers learned that a high protein diet increased milk yields. They used it alongside their traditional breeding programmes to get the best out of their Jerseys or their Holsteins. Scientific underpinning came quite late when experiments on sheep showed how useful it was in supplying amino acids to ruminants. These results boosted MBM in the 1980s. But cannibalism of a sort had been going on much longer. The best indicator is anthrax. This bacterium achieved notoriety in the autumn of 2001 when a bioterrorist in the USA made letters into lethal weapons by filling them with its spores. In Britain at the beginning of the twentieth century about 1000 cattle died from anthrax every year. Much of it came in bones from the Indian subcontinent. There was a flourishing bone import trade; sun-dried ones found in deserts and jungles were particularly desirable because they had a low fat content and could be made into meal or other products directly without heat treatment or solvent extraction. Unfortunately, such bones contain anthrax spores, no doubt because a proportion of them came from animals that had died naturally from the disease. Just as with BSE 80 years later some animals in Britain were infected by eating bone meal and others by eating feeds contaminated during processing and handling.

It would be very wrong to underestimate the suffering of those who have died from vCJD or an *E. coli* O157 infection. Their families have also endured much grief. For them it is small consolation to know that such events played out in public and mediated by mechanisms like the Lardner effect, have driven improvements in public health and food safety. Britain now produces safe beef. Butchers are licensed. Food law enforcers have received more money. The Food Standards Agency is well established. HACCP marches on. But before jumping to the conclusion that BSE and *E. coli* O157 have been uniquely modern tragedies, they must be put into a historical context. In 1942 Graham Wilson (later Sir Graham, but then Professor of Bacteriology at the London School of Hygiene and Tropical Medicine) published a review, *The Pasteurization of Milk*. He had been asked to prepare it by the Minister of Health after a 1938 bill allowing local authorities to make milk pasteurization compulsory had failed in the House of Commons. He estimated that between 1912 and 1937, 65,000 people in England and Wales died from milk-borne bovine tuberculosis. Wilson was very cautious. It is impossible to disagree with his caveat that his estimate erred on the conservative side, or his conclusion that in 1942 'raw milk is probably the most dangerous article in our dietary'. The scandal was not only that these deaths could be prevented by pasteurization (Londoners did not get bovine TB because London milk was pasteurized) but that pasteurization had been prevented, as Dr Edith Summerskill said when introducing a successful pasteurization bill in 1949, by 'ignorance, prejudice, and selfishness'. Anti-pasteurization lobbyists used many arguments. A long running one was the complaint that regulations were being made by town people who did not understand country life. Such attitudes are, of course, still with us.

The twentieth-century tuberculosis death toll fatally undermines the foundations of the 'risk society' paradigm. The Irishman viewing the stinking mass of corruption that was his 1847 'organic' potato crop and then watching his family die of typhus or the rachitic Londoner unsuccessfully gasping for breath in the 1952 smog would be amazed at our arrogance and ignorance in giving credence to the idea that we live in particularly dangerous times. That we do, dramatically demonstrates the two things that helped *E. coli* O157 and BSE to cause so much harm—a failure to learn from history, and a failure to understand science. Wishaw would not have happened if the messages from previous food poisoning outbreaks had been heeded. BSE and vCJD were not caused by too much science. Its underapplication was the problem. Britain is brilliant at winning Nobel Prizes but bad at using its best brains to protect the public. The facts speak for themselves!

But the news is not all bad. *E. coli* O157 infections and vCJD remain rare and are not getting commoner. Nevertheless, they still concern policy makers. For *E. coli* O157 emphasis has shifted to the prevention of infections caused by direct contact with manure, particularly when town people visit the country. The problem is how to protect them from the many millions of tons of manure deposited on the British countryside every year. Risk reduction is not straightforward. Cowpats are vastly commoner than infections—even in a high incidence country like Scotland only about 1 case for every 100,000 tons. So measures have to be proportionate, focusing, for example, on the supervising of young children (most at risk from kidney complications) rather than restricting countryside access for all, FMD style. Proportionality is a big issue for vCJD control as well. Since March 1996 cattle older than 30 months have been banned for sale from human consumption (the Over Thirty Months— OTM—rule). This stops cattle about to develop BSE going into the food chain, because the disease only develops in older animals. OTM animals are bought by the government and incinerated. The scheme costs more than £400 million annually. But the number of cases of BSE has fallen massively since the 1992 peak of 37,056 cases. In 2002 it was 544. Laboratory tests for BSE have also become available. Testing OTM animals to identify those about to develop BSE would cost about £40 million per year. Policy makers have recently reviewed the OTM scheme to assess whether its replacement by testing would be detrimental to public health, and concluded that the time has come for its abandonment.

A particularly good thing to report about these reviews—the Scottish Task Force on *E. coli* O157 (2001) and the Food Standards Agency OTM Rule Review (2002-2003)—is that they show that lessons have been learned from past mistakes. Inclusivity and openness have been their hallmarks. If their formats became the norm, it will be possible to say that good has come out of tragedy.

Further Reading

Chapter 1

Determination by G. L. Cox, Sheriff Principal of Sheriffdom of South Strathclyde, Dumfries and Galloway into the *E.coli* O157 Fatal Accident Inquiry. Airdrie, 14 August 1998.

Judgment of stated case *in causa* Haston, R. B. and Haston, M. A. F. Both trading as R. and M. Haston against Ruxton, L. M. 1996. Linlithgow: Sheriffdom of Lothian and Borders.

Kohl, H. S., Chaudhuri, A. K. R., Todd, W. T. A., Mitchell, A. A. B., and Liddell, K. G. (1994). A severe outbreak of *E.coli* O157 in two psychogeriatric wards. *Journal of Public Health Medicine*, *16*, 11-15.

Reason, J. (1997). *Managing the Risks of Organisational Accidents*. Ashgate, Aldershot.

Chapter 2

O'Mahony, M., Mitchell, E., Gilbert, R. J., Hutchinson, D. N., Begg, N. T., Rodhouse, J. C., and Morris, J. E. (1990). An outbreak of foodborne botulism associated with contaminated hazelnut yoghurt. *Epidemiology and Infection*, *104*, 389-95.

Reason, J. (1997). *Managing the Risks of Organisational Accidents*. Ashgate, Aldershot.

Report of the R.101 Inquiry. Abridged edition 1999. R101 The Airship Disaster 1930. The Stationery Office, London.

The Public Inquiry into the Piper Alpha Disaster. The Hon Lord Cullen. HMSO 1990 Cm 1310. 2 vols.

The Report on the Croydon Typhoid Outbreak (1938). *Public Health*, 51, 179-83. See also The Croydon Typhoid Inquiry (1938). *Public Health*, 51, 137-55.

Snow, J., Richardson, B. W., and Frost, W. H. (1936). *Snow on Cholera being a Reprint of Two Papers by John Snow M. D. together with a Biographical Memoir by B. W. Richardson, M. D. and An Introduction by Wade Hampton Frost, M. D.* The Commonwealth Fund, New York.

Waddington, C. H. (1973). *OR in World War 2. Operational Research against the U boat.* Elek Science, London.

Chapter 3

Gettings, H. S. (1913). Dysentery, Past and Present. *Journal of Mental Science*, 59, 605-21.

Gilbert, R. J. and Maurer, I. M. (1968). The hygiene of slicing machines, carving knives and can-openers. *Journal of Hygiene*, 66, 439-50.

Goffman, E. (1968). *Asylums. Essays on the Social Situation of Mental Patients and Other Inmates.* Penguin, Hardmondsworth.

House of Commons Welsh Affairs Committee Third Report Session 1990-91. Arrangements for Handling Serious Outbreaks of Food Poisoning in the Light of the Salmonella Outbreak in July and August 1989 in North Wales. HMSO 1991.

Review of the Handling of the Salmonella Outbreak in July/August 1989. The Welsh Office 1990.

The Report of the Committee of Inquiry into an Outbreak of Food Poisoning at Stanley Royd Hospital. HMSO, 1986. Cmnd 9716.

The Aberdeen Typhoid Outbreak. Report of the Departmental Committee of Inquiry. 1964. Cmnd 2542. HMSO Edinburgh.

Chapter 4

Determination by G. L. Cox, Sheriff Principal of Sheriffdom of South Strathclyde, Dumfries and Galloway into the *E.coli* O157 Fatal Accident Inquiry. Airdrie, 14 August 1998.

Nightingale, F. (1860) Notes on Nursing: *What It Is And What It Is Not.* Harrison, London.

The Public Inquiry into the Piper Alpha Disaster. The Hon Lord Cullen. HMSO 1990 Cm 1310. 2 vols.

Chapter 5

Campbell, A. W. (1895). The Breaking Strain of the Ribs of the Insane. An analysis of a series of fifty-eight cases tested with an instrument specially devised by Dr C. H. Mercier. *Journal of Mental Science*, 41, 254-74.

Grashey, H von (1886). Bernhard von Gudden†. Nekrolog. *Archiv für Psychiatrie und Nervenkrankheiten*, 17, i–xxix.

Gudden, B (1863). Über die Entstehung der Ohrblutgeschwulst. *Allgemeine Zeitschrift für Psychiatrie*, 20, 423-30.

Gudden, B. (1870). Ueber die Rippenbrüche bei Geisteskranken. *Archiv Fur Psychiatrie und Nervenkrankheiten*, *2*, 682-92.

Hunter, R. and Macalpine, I. (1974). *Psychiatry for the Poor. 1851 Colney Hatch Asylum Friern Hospital 1973*. Dawsons, London.

Ireland, W. W. (1886). The Insanity of King Louis II of Bavaria. *Journal of Mental Science*, *32*, 332-46.

Lardner, D. (ed) (1859). *The Museum of Science and Art*, Vol. 1. Walton and Maberly, London.

McLean, I. and Johnes, M. (2000). *Aberfan, Government and Disasters*. Welsh Academic Press, Cardiff.

Middlemass, J. and Robertson, W. F. (1894). Pathology of the Nervous System in Relation to Mental Diseases. *Edinburgh Medical Journal*, *40*, 509-18.

Norman, F. (1958). *Bang to Rights*. Secker & Warburg, London.

Occasional Notes of the Quarter. The Late Samuel Gaskell, Esq. (1886) *Journal of Mental Science*, *32*, 235.

Reade, C. (n.d.). *Hard Cash*. Collins, London.

Report of the Walkerton Inquiry. Part One. The Queen's Printer for Ontario. 2002.

Reports on the Progress of Psychological Medicine 1870. 2. The Case of Manslaughter at Lancaster Asylum. *Journal of Mental Science*, *16*, 129-33.

Rhodes, G. (1981) *Inspectorates in British Government*. George Allen & Unwin, London.

Rolt, L. T. C. (1966). *Red for Danger. A History of Railway Accidents and Railway Safety Precautions*. Pan, London.

The Report of the Committee of Inquiry into an Outbreak of Food Poisoning at Stanley Royd Hospital. HMSO, 1986. Cmnd 9716.

Stretton, C. E. (1893). *Safe Railway Working*. Crosby Lockwood, London.

The Ladbroke Grove Rail Inquiry. Reports Parts 1 and 2. The Rt. Hon. Lord Cullen. HSE Books, 2001.

Walton, J. (1981). *The Treatment of Pauper Lunatics in Victorian England. The Case of Lancaster Asylum, 1816-1870*, pp. 166-97, in Scull, A. (ed), *Madhouses, Mad-doctors and Madmen*. Athlone Press, London.

Chapter 6

Bell, B. P. *et al.* (1994). A multistate outbreak of *Escherichia coli* O157: H7—associated bloody diarrhoea and haemolytic uraemic syndrome from hamburgers: the Washington experience. *Journal of the Americal Medical Association*, *272*, 1349-53.

Fox, N. (1997). *Spoiled. The Dangerous Truth About a Food Chain Gone Haywire*. Basic Books, New York.

Kaper, J. B. and O'Brien, A. D. (eds) (1998). *Escherichia coli O157:H7 and Other Shiga Toxin-Producing E.coli Strains*. ASM Press, Washington DC.

Morgan, M. G. *et al.* (1988). First recognized community outbreak of haemorrhagic colitis due to Verotoxin-producing *Escherichia coli* O157:H7 in the UK. *Epidemiology and Infection*, *101*, 83-91.

Riley, L. W. *et al.* (1983). Haemorrhagic colitis associated with a rare *Escherichia coli* serotype. *New England Journal of Medicine*, *308*, 681-5.

Strachan, N. J. C., Fenlon, D. R., and Ogden, I. D. (2001). Modelling the vector pathway and infection of humans in an environmental outbreak of *Escherichia coli* O157. *FEMS Microbiology Letters*, *203*, 69-73.

Chapter 7

Cairns, J., Stent, G. S., and Watson, J. D. (eds) (1992). Expanded edn. *Phage and the Origins of Molecular Biology*. Laboratory Press, Cold Spring Harbor.

Cotton, H. A. (1923) The Relation of Chronic Sepsis to the So-called Functional Mental Disorders. *Journal of Mental Science*, *69*, 434-65.

Escherich, T. (1885). Die Darmbacterien des Neugeborenen und Säuglings. *Forschritte der Medezin*, *3*, 515-22, 547-54.

Fischer, E. P. and Lipson, C. (1988). *Thinking About Science. Max Delbruck and the Origins of Molecular Biology*. Norton, New York.

Gemmel, J. F. (1898). *Idiopathic Ulcerative Colitis*. Bailliere, Tindall and Cox, London.

Kaper, J. B. and O'Brien, A. D. (eds) (1998). Escherichia coli *O157:H7 and Other Shiga Toxin-Producing* E.coli *Strains*. ASM Press, Washington DC.

Mill, J. S. (1875). *Nature, The Utility of Religion, and Theism*. 4th edn, Longmans, Green, Reader and Dyer, London.

Mott, F. W. and Durham, H. E. (1900). Report on Colitis or Asylum Dysentery. London.

Neidhardt, F. C. (editor in chief) (1996) 2nd edn, *Escherichia coli* and *Salmonella*. 2 vols. ASM Press, Washington, DC.

Perna, N. T. *et al*. Genome Sequence of Enterohaemorrhagic *Escherichia coli* O157:H7, *Nature*, *409*, 529-33.

Scull, A. (1990). Desperate remedies: a Gothic tale of madness and modern medicine, pp 144-169 in Murray, R. M., Turner, T. H. (eds) *Lectures on the History of Psychiatry. The Squibb Series*. Gaskell, London.

Smith, J. (1955). *The Aetiology of Epidemic Infantile Gastro-enteritis*. The Royal College of Physicians, Edinburgh.

Stahl, F. W. (ed) (2000). *We Can Sleep Later. Alfred D. Hershey and the Origins of Molecular Biology*. Laboratory Press, Cold Spring Harbor.

Chapter 8

Brown, P. (1994). *Transmissible Human Spongiform Encephalopathy (Infectious Cerebral Amyloidosis): Creutzfeld–Jakob Disease, Gerstmann–Straüssler–Scheinker Syndrome, and Kuru*, pp 839-76 in Calne, D. B. (ed), *Neurodegenerative Diseases*. W. B. Saunders, Philadelphia.

Prusiner, S. B. (ed) (1999). *Prion Biology and Diseases*. Laboratory Press, Cold Spring Harbor.

Will, R. G. *et al*. (1996). A new variant of Creutzfeld–Jakob disease in the UK, *Lancet*, *347*, 921-5.

Chapter 9

Alper, T., Cramp, W. A., Haig, D. A., and Clarke, M. C. (1967). Does the agent of scrapie replicate without nucleic acid?, *Nature*, *214*, 764-6.

Cairns, J., Stent, G. S., and Watson, J. D. (eds) (1992). Expanded edn. *Phage and the Origins of Molecular Biology*. Laboratory Press, Cold Spring Harbor.

Creager, A. N. H. (2002). *The Life of a Virus. Tobacco Mosaic Virus as an Experimental Model, 1930–1965.* University of Chicago Press, Chicago.

Dickinson, A. G. (1975). Host-pathogen interactions in scrapie, *Genetics, 79,* 387–95.

Dickinson, A. G. and Meikle, V. M. H. (1971). Host-genotype and agent effects in scrapie incubation: change in allelic interaction with different strains of agent, *Molecular and General Genetics, 112,* 73–9.

Fischer, E. P. and Lipson, C. (1988). *Thinking About Science. Max Delbruck and the Origins of Molecular Biology.* Norton, New York.

Fraser, D. (1967). *Viruses and Molecular Biology.* Macmillan, New York.

Friedman, R. M. (2001). *The Politics of Excellence. Behind the Nobel Prize in Science.* Times Books, New York.

Mill, J. S. (1875). *Nature, The Utility of Religion, and Theism,* 4th edn, Longmans, Green, Reader and Dyer, London.

Pauling, L., and Delbrück, M. (1940). The Nature of Intermolecular Forces Operative in Biological Processes, *Science, 92,* 77–9.

Prusiner, S. (1982). Novel Proteinaceous Infectious Particles Cause Scrapie, *Science, 216,* 136–44.

Schrödinger, E. (1969). *What is Life?* Cambridge University Press, Cambridge.

Sime, R. L. (1997). *Lise Meitner.* University of California Press, Berkeley.

Weissmann, C. (1991). A unified theory of prion propagation, *Nature, 352,* 671–83.

Chapter 10

The Plaintiffs v The United Kingdom Medical Research Council and the Secretary of State for Health. High Court of Justice Queen's Bench Division. Judgment of Morland J, 1996.

Wells, G. A., Scott, A. C., Johnson, C. T., Gunning, R. F., Hancock, R. D., Jeffrey, M., Dawson, M., and Bradley, R. (1987). A novel progressive spongiform encephalopathy in cattle, *Veterinary Record, 121,* 419–20.

Brown, P. *et al.* (2000). Iatrogenic Creutzfeld-Jakob disease at the millennium, *Neurology, 55,* 1075–81.

Daubert v Merrell Dow Pharmaceticals (92-102), 509 US 579 (1993). Opinion of Blackmun J, 1993.

Fleck, L. (1979). *Genesis and Development of a Scientific Fact.* University of Chicago Press, Chicago.

Gowing, M. (1964). *Britain and Atomic Energy, 1939–1945.* Macmillan, London.

Kavanagh, D. and Richards, D. (2001). Departmentalism and Joined-up Government: Back to the Future?, *Parliamentary Affairs, 54,* 1–18.

Marx, G. (1998). Leo Szilárd Centenary Volume. Eötvös Physical Society, Budapest.

Price, D. K. (1962). *Government and Science.* Oxford University Press, New York.

Sagan, S. D. (1993). *The Limits of Safety.* Princeton University Press, Princeton.

Chapter 12

d'Aignaux, J. N., Cousens, S. N., and Smith, P. G. (2001). Predictability of the UK Variant Creutzfeld–Jakob Disease, *Epidemiol Science*, 294, 1729-31.

Fleck, L. (1979). *Genesis and Development of a Scientific Fact*. University of Chicago Press, Chicago.

General Paralysis and its Treatment by Induced Malaria. Board of Control (England and Wales). Report by Meagher, E. T. HMSO, 1929.

Kraepelin, E. (1926). The Problems Presented by General Paralysis, *Journal of Nervous and Mental Disease*, 63, No. 3.

Mercier, C. A. (1914). *A Text-Book of Insanity and Other Mental Diseases*, 2nd edn, George Allen & Unwin, London.

Valleron, A.-J., Boelle, P.-Y., Will, R., and Cesbron, J.-Y. (2001). Estimation of Epidemic Size and Incubation Time Based on Age Characteristics of vCJD in the United Kingdom, *Science*, 294, 1726-8.

Whitrow, M. (1993). *Julius Wagner-Jauregg*. Smith-Gordon, London.

Wile, U. J. (1913). The Demonstration of the *Spirochaeta pallida* in the Brain Substance of Living Paretics, *Journal of the American Medical Association*, 61, 866.

Chapter 13

Communication from the commission on the Precautionary Principle. Commission of the European Communities. Brussels, 2 February 200. Com (2000) 1.

Harremoës, P., Gee, D., MacGarvin, M., Stirling, A., Keys, J., Wynne, B., and Guedes Vas, S. (2002). *The Precautionary Principle in the 20th Century*. EARTHSCAN, London.

The Queen v Ministry of Agriculture, Fisheries and Food, Commissioners of Customs and Exise, ex-parte: National Farmers Union et al., Case C-157/96. Court of Justice of the European Communities. Judgment. 1997.

Snow, J., Richardson, B. W., and Frost, W. H. (1936). *Snow on Cholera being a Reprint of Two Papers by John Snow M. D. together with a Biographical Memoir by B. W. Richardson, M. D. and An Introduction by Wade Hampton Frost, M. D.* The Commonwealth Fund, New York.

Neustadt, R. E. and Fineberg. H. (1983). *The Epidemic That Never Was. Policy-Making And The Swine Flu Affair*. Vintage Books, New York.

Chapter 14

Arnold, L. (1992). *Windscale 1957. Anatomy of a Nuclear Accident*. Macmillan, Basingstoke.

Beck, U. (1992). *Risk Society. Towards a New Modernity*. Sage, London.

Douglas, M. and Wildavsky, A. (1983). *Risk and Culture*. University of California Press, Berkeley.

Foot and Mouth Disease 2001: Lessons to be Learned Inquiry Report. London: The Stationery Office 2002 HC888.

Leighton, G. (1923) *Botulism and Food Preservation (The Loch Maree Tragedy)*. Collins, Glasgow.

O'Neill, O. (2002). *A Question of Trust. The BBC Reith Lectures 2002.* Cambridge University Press, Cambridge.

Persey and Others v Secretary of State for Environment, Food and Rural Affairs and Others. England and Wales High Court (Administrative Court) Judgment of Simon Brown LJ. and Scott Baker J. 15 March 2002.

R v Secretary of State for Health ex parte Wagstaff and *R v Secretary of Statement for Health ex parte Associated Newspapers Ltd.* Supreme Court of Judicature Divisional Court, London. Judgment of Kennedy LJ., 20 July 2000.

Report of the Committee appointed to inquire into the circumstances, including the production, which led to the use of contaminated infusion fluid in the Devenport section of Plymouth General Hospital. HMSO, 1972. Cmnd 5035.

Rigden, J. S. (1987). *Rabi Scientist and Citizen.* Basic Books, New York.

Report of the President's Commission on the Accident at Three Mile Island. US Government Printing Office. 1979.

Report of the Committee of Inquiry into the Smallpox Outbreak in London in March and April 1973. HMSO, 1974. Cmnd 5626.

Report of the Committee of Inquiry into Complaints about Ashworth Hospital. Vol. 1 HMSO, 1992, Cm 2028-1.

Report, evidence, and the supporting papers of the Inquiry into the emergence and identification of Bovine Spongiform Encephalopathy (BSE) and variant Creutzfeld–Jakob disease (vCJD) and the action taken in response to it up to 20 March 1996. London. The Stationery Office. 2000. 887. (Vols 1-16 of the Report, written witness statements and transcripts of oral hearings were used in the preparation of this book.)

Royal Commission on Tribunals of Inquiry 1966. Report of the commission under the chairmanship of the Rt. Hon. Lord Justice Salmon. HMSO Cmnd 3121.

Report of the Committee of Inquiry on Anthrax. HMSO, 1959. Cmnd 846.

Report by R. Moelwyn Hughes upon the Enquiry into the Disaster at the Bolton Wanderers Football Ground on the 9th March, 1946. HMSO. 1946. Cmd 6846.

Sills, D. L., Wolf, C. P., and Shelanski, V. B. (1982). *Accident at Three Mile Island: The Human Dimensions.* Westview, Boulder, Colorado.

The Pennington Group Report of the Circumstances leading to the 1996 Outbreak of Infection with *E.coli* O157 in Scotland, the implications for Food Safety, and the Lessons to be Learned. 1997. The Stationery Office, Edinburgh.

Wilson, G. S. (1942) *The Pasteurization of Milk.* Edward Arnold, London.

Index

Aberdeen Typhoid outbreak 42–45
Aberfan disaster 66–68
Acheson, Sir Donald 155, 178, 200
Advisory Committee on the Microbial
    Safety of Food 202
Ahmed, Syed 3, 4, 13
Allen, Ingrid 118
Alper, Tikvah 129, 130
Alzheimer, Alois 121, 188
amateurs 157, 178
Anderson, Sir John 177–178
Anthrax 212
Arber, Werner 114
Ashworth Hospital 78–79
asylum ear 74–75, 77
asylum dysentery 104–107
atom bomb 175–177, 178, 179

Barr, John 3–10, 12–15, 18–21, 24, 25, 29,
    33, 35, 38, 41, 42, 58, 59, 81, 82
Beck, Ulrich
    "Risk Society" 211
beef, corned 19, 43, 55
beef, roast 7, 9, 10, 11, 19, 50
Beijerinck, Martinus 127, 128
Bethlem Hospital 68–69
Bentham, Jeremy 45
biodegradable detergent 6–7, 20
Blackmun J. opinion in "Daubert" 178–189
Blair, Tony 208
Blom-Cooper, Louis see Inquiries
Bryceland, Graham 3, 4, 12, 13, 61

BSE
    and bulls eyes 166–167, 173
    and cats 164
    and exotic species 165
    and John Wilesmith 147
    and meat and bone meal : see meat and
        bone meal
    and mechanically recovered meat 164, 171
    and pet food 163
    and rendering 148
    and scrapie 143, 144, 146, 147, 148, 149,
        154, 155, 162, 174–175, 196
    and spinal cord 167
    and tripe 163
    December 1986 minute by
        Ray Bradley 144
    first description 143
    infectivity of meat and bone meal 175
    Pitsham Farm cases 154
    product of evolution 179
    publication of paper by Gerald Wells 147
    transmission to humans "remote" 160,
        166, 167, 170, 178, 183–184, 201
    Truro case 1987 145
    typing 154
    withdrawal of permission to
        publish 146
BSE Inquiry 165, 199–201, 204, 208, 210
    criticism of individuals 200
    lessons learned 199–200
Bohr, Niels 112, 131, 176
"botox" 28–29

botulism
  Hazelnut yoghurt outbreak 28–29, 41, 201
  Loch Maree outbreak 201
brain biopsy 191
Bridgman, June 199
Broad Street Pump 38–40
Brown, Fred 157, 175
Bruce, Moira 136

Cabinet Ministerial Group on
      Food Safety 27, 161
Calman, Sir Kenneth 200
Campylobacter 89
Carter, President Jimmy 195
Centers for Disease Control, Atlanta 96–97
Central Scotland *E.coli* O157
      outbreak – also *see* Inquiries
  Bankview Nursing Home, Bonnybridge
      9, 11, 17, 37
  Cascade Public House 7–8, 16, 17, 18,
      37, 41
  clinical effects 17–18
  cross contamination 20, 41–42
  E.coli O157 in Barr's premises 19
  establishment of Outbreak Control
      Team 3–4
  HOLMES 15–16
  killing of outbreak strain of *E.coli* O157
      by heat 23
  Law Hospital 1, 2, 8, 13
  Monklands General Hospital 1, 3, 5, 17
  Scottish Centre for Environmental
      Health 14
  Scottish Reference Laboratory 14–15
  start, 2
  Wishaw Parish Church 9–12, 16, 19,
      23–24, 37
Central Veterinary Laboratory 143–147,
      154, 179
"Challenger" space shuttle disaster 180–184
Chandler, Dick 126, 130
Chase, Martha 112
Churchill, Sir Winston 177, 190
CJD
  amyloid plaques 121, 122, 140
  different kinds 118
  Gerstmann–Sträussler–Scheinker
      syndrome 121, 122
  iatrogenic and growth hormone – *see*
      Dickinson Alan
  in Australia and New Zealand 149
  in young people 118
  kuru, and cannibalism 119
  spongiform degeneration 122
  Surveillance Unit, Edinburgh 117–119

  transmissibility 119
Clarke, Kenneth 21, 161
Cockroaches 3, 50
Collinge, John 170
Colney Hatch Asylum 70–71
Cox, Sheriff Graham 20–21, 23, 24, 61, 64
Creutzfeld, Hans Gerhardt 120, 121
cross-contamination 6, 9, 20–21, 41,
      43–45, 55–56
Crown immunity 81–82
Croydon typhoid outbreak *see* Outbreaks
Cullen, Lord 62–64, 82, 90
Currie, Edwina 26, 160, 169

Daubert case 178–179, 180
Davidson, Mrs 9, 10–11
Dealler, Stephen 173
Delbrück, Max 112–113, 129–132, 139,
      140–141
Department of Energy, Safety Directorate
      of Petroleum Energy Division
      62–64, 71, 82
Department of Health, London, 56, 117,
      152, 157, 159–160, 166–171, 202, 208
departmentalism 173
"detoxication" 107
de Valera, Eamonn 129
Devine, John 22, 58–59
Dickens, Charles 54, 70
Dickinson, Alan
  and scrapie genetics 133–134
  and *sinc* gene 134–135
  and the Neuropathogenesis Unit,
      Edinburgh 149
  and typing of scrapie 135–136
  "battle weariness" 156
  his "virino" hypothesis 140
  starts work on scrapie 132
  thought processes 133
  warns about possible transmission of CJD
      by growth hormone 133, 150–154
Dorrell, Stephen 171, 183, 201

E.coli
  and sex 104
  bacteriophages 111–112
  clones 105
  enteropathogenic 109–110
  enteropathogenic in Aberdeen
      outbreak 109
  evolution 110
  genetics 111–114
  genome sequence 110
  Hershey–Chase experiment 112, 127
  K12 110

restriction enzymes 114
E.coli O157
  and McDonalds 97
  clonal nature 104
  genome sequence 110
  haemolytic uraemic syndrome 4, 17, 91,
    97, 98, 99, 100
  how it causes disease 99–100
  infectious dose 100–102
  novelty 96
  origin and evolution 103
  outbreaks *see* Outbreaks
  Task Force 213
  toxins 99, 104
  thrombotic thrombocytopaenic purpura
    17, 99, 100
  UK outbreaks 98
  US outbreaks 98
Edinburgh Royal Infirmary 107
Electron microscope and negative staining
  128, 120
Epstein, (Sir) Antony 156, 157
Escherich, Theodore 103–104

Fatal Accident Inquiries *see* Inquiries
Ferguson–Smith, Malcolm 199
Feynman, Richard 183
Fineberg, Harvey *see* Neustadt, Richard E
Fleck, Ludwik
  career 190–191
  exoteric and esoteric circles 172–173
  "Genesis and Development of a Scientific
    Fact" 172
  in Auschwitz and Buchenwald 190
  on syphilis 189
  thought collective 172, 174, 180
  thought style 172–173
  work on typhus 190
Food Safety Act 26, 27, 59, 82
Food Standards Agency 208, 212–213
Foot and Mouth Disease 208–210
Ford, President Gerald 194
Forsyth, Michael 15, 201, 204
Fraënkel–Conrat, Heinz 127
Fraser, Hugh 136
French Revolution 208
Frisch, Otto 176, 200
Frisch–Peierls memorandum 176, 179

Gadjusek, Carlton 119
"Garbage Can" model of decision
  making 174
Gardner, Richard S. (author's grandfather)
  73, 78, 87–88
Gaskell, Samuel 70, 72, 77, 95

Gemmel, J.F. 105–106
General paralysis of the insane 186–189
Gerstmann, Josef 121
Gestapo 113, 120
Gilbert, Richard 45
Goffman, Erving 53
Gordon, W.S. 125
Gowing, Margaret 176
Gudden, Bernhard von 75–77, 79, 122, 188
Gummer, Cordelia 165, 196
Gummer, John 165–167

Hazard Analysis and Critical Control
  Points (HACCP) 202–203, 206
Her Majesty the Queen 43
Hershey, Alfred 112, 113
Hine, Dierdre 167–169, 173
Hogg, Douglas 171

Inspectorates
  anatomy 66
  efficiency 65
  enforcement 65–66
  factories 66
  lunacy 66, 69, 77, 78–80, 105
  mines and quarries 66, 68
  railways 66, 82–88
Inquiries
  Aberdeen typhoid 44
  Aberfan 68
  Ashworth Hospital 78–79
  Blom–Cooper's criteria for success
    204–208
  BSE *see* BSE Inquiry
  Central Scotland *E.coli* O157 6–7, 16, 29,
    20–21, 22, 23–24, 29, 37, 58, 60–61,
    64, 201, 204
  "Challenger" 183
  cost 200
  Croydon typhoid 30–31
  different kinds 198
  Fatal Accident, description 16
  Hartwood Hill Hospital 2–3
  "Herald of Free Enterprise" 210
  Ladbroke Grove 90
  Loch Maree 201
  Marconi 198
  North Wales–Cheshire *Salmonella* 56–57
  Pennington 15, 201–208
  Piper Alpha 35, 62–64
  public 199
  R101 airship 36
  Shipman 210–211
  Stanley Royd Hospital 49–55, 82
  Windscale Fire 1957 205

Irish famine 213
Ivanovskii Dimitri 127, 128

James, Philip 208
Jacob, Francois 113–114
Jakob, Alfons 121

Kimberlin, Richard 146, 151, 156, 163
Koebel, Frank 92–93
Koebel, Stan 92–93

Lacey, Richard 173
Lancaster Asylum 70, 72–73, 88, 95,
    105–106, 187
  general paralysis of the insane in, 189
  Refractory Ward 72, 73, 187
  William Wilson (patient) 72–73, 79, 187
Lardner, Dionysius 86–90
Lardner effect 89
Law Hospital 1, 2, 8, 13
Lederberg, Joshua 110, 113
"leaning into the wind" 200
licensing of butchers 203, 206–207, 212
licensing of football grounds 207–208
Liddell, Kenneth 2, 3, 9
London Smog 1952 213
Louping ill 125
Ludwig II of Bavaria 75–77
Luria, Salvador 112, 113
Lwoff, André 113

Macdonald, Ramsay 35
MacGregor, John 155, 158, 160, 193
MacQueen, Ian 43
Major, John 170, 171, 201
"*mala prohibita*" 82
MAUD Committee 176–177, 179, 180
McLean, Drew 5, 8, 21
meat and bone meal 147–149, 154, 160, 163,
    164, 175, 211–212
Meat Products (Hygiene) Regulations 1994
    21–22, 59
Meitner, Lise 130, 131, 132, 176, 180, 200
Meldrum, Keith 164–165
Mercier, Charles 80, 187
Mendel, Gregor 111, 131
Milk 3–4, 98, 212
Mill, John Stuart on Nature 107–108
Ministry of Agriculture, Fisheries and Food
    (MAFF) 26, 149, 155, 157–160, 163,
    164, 166, 170, 171, 174, 175, 200, 202,
    208, 210
Monklands General Hospital 1, 3, 5, 17
Monod, Jacques 112–114

Moon, David 7–8, 41
Montagu, Lord 155, 158, 174

Nathans, Daniel 114
Neustadt, Richard E and Fineberg, Harvey
    "The Swine Flu Affair: Decision-Making
    on a Slippery Disease" 195–196
Nightingale, Florence 62, 80
Norris, James 68–69
North Lanarkshire Environmental Health
    Department 3–4, 5, 7, 8, 13, 25, 42,
    59, 61–62
North Wales–Cheshire *Salmonella* outbreak
    1989 55–57

O'Neill, Onora
  "A Question of Trust" 208
Oppenheimer, J. Robert 175–176
Outbreaks
  Aberdeen enteropathogenic *E.coli* 109
  Aberdeen, typhoid 42–45
  Central Scotland *E.coli* O157 : *see*
    Central Scotland
  Cholera, Broad Street 38–40
  Croydon, typhoid 29–30, 32, 35,
    36–37, 41
  East Anglia *E.coli* O157 98
  Foot and Mouth disease 2001 208–210
  Hartwoodhill Hospital *E.coli* O157 2–3,
    9, 106
  Hazelnut yoghurt, botulism 28–29, 41
  Jack-in-the-Box *E.coli* O157 97–98
  Loch Maree, botulism 201
  Medford, Oregon *E.coli* O157 97
  New Deer *E.coli* O157 100–101
  North Wales-Cheshire, *Salmonella* 1989
    55–57
  Sakai City *E.coli* O157 98–99
  Stanley Royd, *Salmonella* 46–51
  Swine influenza US 1976 193–196
  Traverse City *E.coli* O157 97
  Walkerton *E.coli* O157 91–94
  West Lothian *E.coli* O157 4
Over Thirty Month Rule 213

pasteurisation 4, 26, 212
Pattison (Sir) John 170
Peierls, Rudolph 131, 176–177
Phillips, Lord 165, 199
Pinderfields Hospital 51–52, 94
Piper Alpha disaster 31–35, 36, 37, 41,
    62–64, 83
Pickles, Hilary
  description of 157, 169

drafting Southwood Report 159–160
letter to Deirdre Hine 167–169
Plurenden Manor Farm 143
precautionary principle
 a bad guide 193
 and foot and mouth disease 209–210
 definition 193
 European Commission
  "Communication" 196–197
 legal use 193
 not straightforward for BSE 209
Price, Don K
 "Government and Science" 173–174
PrP 123
PrP^{sc} 122–123, 139–141
PrP knockout mice 136
Proctor, Richard 59–61
Prusiner, Stanley
 his "prion" hypothesis 137–140
 identifies PrP and PrP^{sc} 136
 purification of scrapie 136
 works on *E.coli* 136
Public Health Laboratory Service 96,
  167–168, 170, 174
Purdey, Mark 205

Railways
 and Dionysius Lardner 89
 Armagh accident 85–86
 couplings automatic and
  non-automatic 87
 Ladbroke Grove accident 90
 Quintinshill accident 83–84
 Regulation Act 86
 Royal Commission on Accidents to
  Railway Servants 86
Reade, Charles
 'Hard Cash' 72
Reagan, President Ronald 182
Reason, James 27
 "resident" and "latent" pathogens
  19–20, 50
"remoteness" *see* BSE
Rooke, Stephen 13, 14, 205
R101 airship disaster 35–36, 83

Sabin, Albert 194
Sagan, Scott "The Limits of Safety" 174
Salmonella 26–27, 49–50, 52, 55–56, 81,
  103, 160
Salmonella typhi 43
Salk, Jonas 194
Schramm, Gerhardt 127
Schrödinger, Erwin 129, 131

Scrapie
 description of disease 124–125
 eradication attempts in US 126
 history 124
 hypothesis and BSE *see* BSE and scrapie
 in cattle 175
 in Europe 124
 in mice 126, 130
 irradiation 129
 transmissibility 125
"Scrapie hypothesis" *see* BSE
Scotmid 8, 9, 12, 18, 19
Scottish Office 13, 14, 201–203
Sencer, David 194–196
Shiga, Kyoshi 104
Shigella dysenteriae 104
Shigella flexneri 104
Shigella sonnei 104
Smith, Hamilton O 114
Smith, John 109
Snow, John 38–40, 149
Southwood, Sir Richard
 appointment as Working Party Chairman
  155–156, 174
 career 155
 on Editorial Board of Entomologist's
  Monthly Magazine 156
 role 178
 working party meetings 157–159
Southwood Report (also *see* Southwood,
  Sir Richard; and Pickles, Hilary)
 and a Fleck esoteric circle 173
 and baby food 161–162
 and precautionary principle 193
 conclusions 160–161, 174–175
 reception by Cabinet 161–162
Spongiform Encephalopathy Advisory
  Committee (SEAC) 118, 119, 170,
  171, 173
Stanley Royd Hospital
 and Sir David Ferrier 122
 asylum ear at 74
 diarrhoea at 45–47
 history 45
 kitchens 47–48, 50–55, 80
 Salmonella outbreak 46–51
steak pie 9–12
syphilis in Aberdeen in 1497 197
Szilard, Leo 180, 200

Tatum, Edward 110, 113
Thatcher, Margaret 26, 161–162
Three Mile Island nuclear reactor accident
  204–205

Tobacco mosaic virus
  filtration 127
  infectivity of RNA 127
Tonner, Jeffrey 3–6, 8–9, 20–21, 37
tuberculosis 212–213
typhus fever 190–191

ultracentrifuge 128–129

Vaughan, Diane
  "The Challenger Launch Decision" 183
vCJD
  and Gulcan Hassan 116–117, 118
  and Stephen Churchill 115–116, 118
  and tonsils and appendixes 186
  brain biopsy 191
  excess of cases in northern Britain 186
  in Leicestershire 186
  palliative care 187
  prediction of case numbers 185–186
  tests for 191

Wagner–Jauregg, Julius 188–189
Weigl, Rudolf 190
Weissman, Charles 137, 141–142
Wells, Gerald 143, 145, 146, 147
Welsh Office 56, 167–168
Wildy, Peter 152–153
Wilesmith, John 147, 154, 155,
    157, 174
Will, Robert 117, 174
Windscale nuclear reactor accident 205
World Trade Center 37
Wynne–Williams (Flint) Ltd 55–56